Ancient Water Technologies

Ancient Water Technologies

Larry W. Mays
Editor

Springer

Editor
Larry W. Mays
School of Sustainable Engineering and the
Built Environment
Arizona State University
Tempe, AZ 85287-5306
USA
mays@asu.edu

ISBN 978-90-481-8631-0 e-ISBN 978-90-481-8632-7
DOI 10.1007/978-90-481-8632-7
Springer Dordrecht Heidelberg London New York

Library of Congress Control Number: 2010922811

Cover illustration: Aqueduct near Milas, Turkey for supplying water to ancient Mylasa, photo by Larry W. Mays, taken May 2009.

Printed on acid-free paper

Springer is part of Springer Science+Business Media (www.springer.com)

Preface

Water technology began during antiquity long before the great works of investigators such as Leonardo da Vinci (1452–1519) and Daniel Bernoulli (1700–1782). The history of water technology started even before Archimedes (287–212 B.C.).

Moreover, great water projects were already built thousands of years before the development of the concepts of conservation of mass, energy and momentum, (which are used in present-day water project designs) even existed.

This book presents an introduction to ancient water technology. It is different from other books related to ancient water technologies and concentrated on specific ancient civilizations, in that it presents a more universal picture of ancient water technology. It is written by authors from multidisciplinary fields ranging from engineering, water resources engineering, hydrology to archaeology, architecture and geology.

The entire spectrum of ancient water technologies can never be covered in one book, let alone by one author, however, this volume provides an excellent overview of the water technologies of many ancient civilizations. These include the very earliest civilizations such as the Mesopotamians and the Indus Valley Civilization, later civilizations such as the Mycenaeans, the Minoans, the Persians, and the Egyptians, followed by water technologies of the Greeks, the Romans, the Urartians, and the Nabataeans. Furthermore, water technologies of ancient civilizations in the Americas, including the Hohokams, the Anasazis, the Teotihuacans, the Xochicalcoans, the Mayans, the Aztecs, and the Incas are also covered.

Each of the chapters presents a detailed discussion on various topics, one can read for example about ancient Greek Lavatories, an analysis of the water system of a Roman city, effects on groundwater resources from earthquakes in antiquity or, the water management of a complex in ancient Iran.

This book has grown out of a sincere passion to learn about water technology developed by ancient civilizations. This has driven me to visit many ancient locations, particularly in Italy, France, Greece, Spain, and Turkey to study and photograph remains of the ancient water systems. Combined with this passion is my interest in sustainability issues, in particular water resources sustainability, and how we may use technologies (traditional knowledge) developed by the ancients to help alleviate and solve some of our present day water resources problems.

Water resources sustainability is the ability to use water in sufficient quantities and quality from the local to the global scale to meet the needs of humans and ecosystems for the present and the future to sustain life, and to protect humans from the damages brought about by natural and human-caused disasters that affect sustaining life. The success and/or failure of the ancient civilizations depended upon their awareness and ability to work with water resources sustainability issues. Our present day future also depends upon similar issues related to water resources sustainability. Studying ancient water technologies may provide answers for our future.

A book is a companion along the pathway of learning. Have a good journey.

Tempe, Arizona Larry W. Mays

Contents

Contributors

Georgios Antonio Architect Engineer, Hellas, 44 Ioannou Soutsou, 11474 Athens, Greece, antonioug@tee.gr

Siamak Arianfar Department of History, Islamic Azad University, Sharhr-Ray Branch, Tehran, Iran, sarianfar@yahoo.com

Andrew Colombo Department of Civil Engineering, University of Toronto, 35 St. George St., Toronto, ON, Canada, andrew.colombo@utoronto.ca

Mahdi Moradi-Jalal Department of Civil Engineering, University of Toronto, 35 St. George St., Toronto, ON, Canada, mahdi@optimal-solution.ca

Yuri Gorokhovich Department of Environmental, Geographic, and Geological Sciences, Lehman College, CUNY, 250 Bedford Park Blvd., West Bronx, NY 10468-1589, USA, yuri.gorokhovich@lehman.cuny.edu

Benoît Haut Université Libre de Bruxelles, Transfers, Interfaces and Processes (TIPs), Chemical Engineering Unit, Av. F.D. Roosevelt 50, C.P. 165/67, 1050 Brussels, Belgium, bhaut@ulb.ac.be

Bryan Karney Department of Civil Engineering, University of Toronto, 35 St. George St., Toronto, ON, Canada, karney@ecf.utoronto.ca

Demetris Koutsoyiannis Department of Water Resources, School of Civil Engineering, National Technical University of Athens, Heroon Polytechneiou 5, GR 15780 Zographou, Greece, dk@itia.ntua.gr

Nikos Mamassis Department of Water Resources, School of Civil Engineering, National Technical University of Athen, Heroon Polytechneiou 5, GR 15780 Zographou, Greece, nikos@itia.utua.gr

Larry W. Mays School of Sustainable Engineering and the Built Environment, Arizona State University, 85287-5306 Tempe, AZ, USA, mays@asu.edu

Aldo Tamburrino Department of Civil Engineering, University of Chile, Casilla 228-3, Santiago, Chile, atamburr@ing.uchile.cl

Lee Ullmann Department of Art History and Archaeology, Columbia University, 1190 Amsterdam Avenue, New York, NY 1002, USA, lzu1@columbia.edu

Didier Viviers Université Libre de Bruxelles, Archaeological Research Centre (CReA), Av. F.D. Roosevelt 50, C.P. 175/01, 1050 Brussels, Belgium, dviviers@ulb.ac.be

About the Editor

Larry W. Mays is a professor in the School of Sustainable Engineering and the Built Environment at Arizona State University, Tempe, Arizona and former chair of the Department of Civil and Environmental Engineering. He has been the Director of the Center for Research in Water Resources at the University of Texas at Austin, where he also held an endowed professorship. He was educated at the University of Missouri at Rolla where he received the B.S. and M.S. degrees in civil engineering and then at the University of Illinois at Champaign-Urbana where he received the Ph.D. degree.

Professor Mays has published extensively in engineering literature and in the proceedings of national and international conferences, and invited chapters in books. In addition he has been the author, co-author, and editor-in-chief of many books. His books include: author of *Water Resources Engineering* and co-author of *Groundwater Hydrology* (both published by John Wiley & Sons, Inc.); author of *Optimal Control of Hydrosystems* (published by Marcel Dekker); and co-author of *Applied Hydrology* and *Hydrosystems Engineering and Management* (both published by McGraw-Hill). He was editor-in-chief of *Water Resources Handbook, Water Distribution Systems Handbook, Urban Water Supply Management Tools, Stormwater Collection Systems Design Handbook, Urban Water Supply Handbook, Urban Stormwater Management Tools, Hydraulic Design Handbook, Water Resources Systems Management Tools, Water Supply Systems Security*, and *Water Resources Sustainability*, all published by McGraw-Hill. In addition, he was editor-in-chief of *Reliability Analysis of Water Distribution Systems* and co-editor of *Computer Methods of Free Surface and Pressurized Flow.* Professor Mays' most recent book is *Urban Water Management in Arid and Semi-arid Regions*, as editor-in-chief, which was published by Taylor and Francis. This book was the result of volunteer work for the United Nations UNESCO-IHP in Paris.

Among his honors is a distinguished alumnus award from the University of Illinois at Champaign-Urbana and a Diplomate, Water Resources Engineering of the American Academy of Water Resources Engineering. He is also a Fellow of the American Society of Civil Engineers and the International Water Resources Association. As registered professional engineer in several states, and a registered professional hydrologist, he has served as a consultant to many organizations. His most recent major research efforts have been in the study of ancient water technologies and the relation that this traditional knowledge could have on solving our problems of water resources sustainability, not only for the present but the future. He enjoys the outdoors, hiking, fishing, alpine skiing, and photography of ancient water systems. Professor Mays lives in Mesa, Arizona and Pagosa Springs, Colorado.

Chapter 1
A Brief History of Water Technology During Antiquity: Before the Romans

Larry W. Mays

1.1 Introduction

Hydraulic technology began during antiquity long before the great works of such investigators such as Leonardo da Vinci (1452–1519), Galileo Galilei (1564–1642), Evangelista Torricelli (1608–1647), Blaise Pascal (1623–1662), Isaac Newton (1642–1727), Daniel Bernoulli (1700–1782), and Leonhard Euler (1707–1783). The history of hydraulic technology even began long before Archimedes (287–212 B.C.). It is amazing to see what was accomplished in the application of water technology during antiquity, millenniums before the development of the concepts of conservation of mass, energy, and momentum used in present-day hydraulic design. Humans have spent most of their history as hunting and food gathering beings. Only in the last 9,000–10,000 years they discovered how to grow crops and tame animals.

The first developments of hydraulics occurred with the development of agriculture using irrigation, followed by the development of urban centers. Brief histories of urban water supply and hydraulic technology in antiquity are provided respectively by Mays et al. (2007) and Mays (2008). This chapter and Chapter 7 provide more detailed discussions of the histories. Mays (2006) provided a discussion on water sustainability and parallels of past civilizations and the present.

1.2 Hydraulic Technology for Irrigation

1.2.1 The Mesopotamians

About 6,000–7,000 years ago, farming villages of the Near East and Middle East became urban centers. During the Neolithic age (ca. 5700–2800 B.C.), the first successful efforts to control the flow of water were driven by agricultural needs (irrigation). Irrigation probably began to develop at a small scale during

L.W. Mays (✉)
School of Sustainable Engineering and the Built Environment, Arizona State University, Tempe, AZ, 85287-5306, USA
e-mail: mays@asu.edu

L.W. Mays (ed.), *Ancient Water Technologies*, DOI 10.1007/978-90-481-8632-7_1,

the Neolithic age in the so-called "fertile crescent," an arc constituting the comparatively fertile regions of Mesopotamia and the Levant, delimited by the dry climate of the Syrian Desert to the south and the Anatolian highlands to the north (Wikipedia). Larger scale irrigation developed when early cultivators settled in the low plains where the Tigris and Euphrates Rivers join. The large alluvial plain between the Tigris and Euphrates Rivers has been occupied by humans since around 5000 B.C., with the beginning of the Sumerian civilization. The large scale diversion of water by humans probably had its origin in ancient Mesopotamia.

The beginning of agriculture by the ancient Egyptians was in the Nile River valley with the original agricultural sites (ca. 5200–4000 B.C.) in the Faiyum depression. Being one of the most predictable rivers in the world, the Nile River flood in Egypt is seldom sudden or abrupt and is timely, in contrast to the floods of the Tigris and Euphrates Rivers. The Nile River Valley in Egypt is a seasonally inundated floodplain during the summer followed by natural drainage. Ideally the Nile would rise to bank-full stage by mid August in southern Egypt, with the northern most basins being flooded four to six weeks later. In contrast the Tigris and Euphrates floods occurred in April or May, which was too early for the fall planting because the summers were too hot. Also the ancient Mesopotamian valley lacked good drainage and experienced many floods. The consequence was that the ancient Mesopotamians had to build canals to divert water from the rivers, to develop their canal irrigated agriculture. Sedimentation in many of the canals was such a critical problem that it was easier to abandon these canals and build new ones. Water technology in ancient Egypt is discussed further in Chapter 3.

Other hydraulic technologies of the Mesopotamians included water tunnels, horse or donkey-powered chain lifts, and at least one major irrigation diversion dam. The ancient diversion dam, the Nimrud Dam, was built across the Tigris River, about 180 km upstream from Bagdad (Butler, 1960). The river water was diverted through the Nahrawan Canal to irrigate an area extending over 100 km to the present town of Baquba. At Baquba this canal joined with the Diyala River to supplement its discharge for the 200 km reach of the Diyala in the ancient course from Baquba to the Tigris River at the modern day city of Kut (Butler, 1960). The ancient diversion dam and the canal not only irrigated the desert area but also transferred water from one river to another.

1.2.2 Lift Irrigation

The first devices for lifting water from a source such as a well or a river or a canal were simple devices that made use of human strength. The *shaduf* had a bag and rope attached to the one end of a wooded arm or beam with a counter balance at the other end of the arm (see Fig. 1.1). The beam rotates around an axis so that the person operating the *shaduf* pulls down the bag into the water, then lifts the bag with water, and then drops the water from the bag. The *shaduf* was known in Mesopotamia as early as the time of Sargon of Akkad (ca. 2300 B.C.). According to the legend

Fig. 1.1 Shaduf (copyright permission with Bruce Salters, used with permission)

about the origin of Sargon, he was supposed to have been abandoned by his mother as a baby into a canal (*naru*) and to have been taken up by a man with a bag of water from this canal. Also there is a shaduf represented on a cylindrical seal from Mesopotamia dated by 2200 B.C. (Viollet, 2006). This technology appeared later in Egypt, with the technology well under way during the 18th Dynasty (ca. 1570 B.C.) and was effective by Roman times (Butzer, 1976). This device allowed the irrigation of crops near the river banks and canals during the summer. Another device for lifting water was the saqiya illustrated in Fig. 1.2. This device is further described in Chapter 3 Water Technology in Ancient Egypt.

1.2.3 Persia and the Qanats

The qanat is a collection and conveyance system for groundwater that was developed in Persia. A qanat, illustrated in Fig. 1.3 consists of an underground tunnel which uses gravity to convey water from the water table (or springs) at higher elevations to the surface of lower lands. Qanats also have a series of vertical shafts that were used

Fig. 1.2 Compartmented wheel with saqiya gear (Description de L Egypte, 1822)

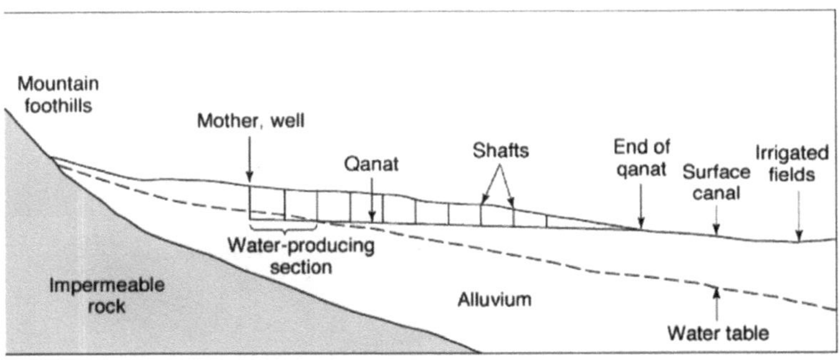

Fig. 1.3 Qanat, foggara, falaj, or karez

for excavation of the tunnel and provided air circulation and lighting. The oldest qanats have been found in the northern part of Iran and date back to around 3,000 years ago when the Arians (Aryans) settled in present day Iran (Javan et al., 2006). The longest (71 km with 2,115 vertical shafts) and oldest (over 3,000 years) is to the ancient city of Zarch. Qanat comes from the Semitic word meaning "to dig" (Moosavi, 2006). Presently there are about 33,000 operational qanats in Iran (Javan et al., 2006).

From 550–331 B.C. Persian rule extended from the Indus to the Nile, during which time qanat technology spread. As this technology transferred to other civilizations, it was known by different names: karez (Afghanistan and Pakistan), kanerjing (China), falaj (United Arab Emirates), and foggara and fughara (North Africa). Qanats were constructed to the west of Persia from Mesopotamia to the Mediterranean and southward into parts of Egypt. Qanats were also constructed to the east of Persia in Afghanistan, in the Silk Route oases settlements of central Asia and to Chinese Turkistan. The Persians introduced qanats to Egypt around 500 B.C.

1.3 Hydraulic Technology for Early Urban Centers

1.3.1 Indus Valley Civilization

One of the early Bronze Age civilizations was Mohenjo-Daro (Mound of the Dead) one of the major urban centers of the Harappa Culture or Indus Civilization. Located on the right bank of the Indus about 400 km south of Karachi, Pakistan, Mohenjo-Daro was a deliberately planned city built around 2450 B.C. over a relatively short time period (Jansen, 1989). This planned city, located in a semi-arid environment, was serviced by at least 700 wells, with an average frequency of one in every third house (Jansen, 1989). The cylindrical well shafts were constructed using wedge-shaped bricks. Probably these circular brick-lined wells were invented by the Harappan people of the alluvial plain. Possibly the wells were built for security purposes in the event of a siege.

Water consumption was large indicated by the bathing platforms in almost every house and the density of the effluent network (Jansen, 1989). The Great Bath of Mohenjo-Daro measured 52 m north-south by 32.4 m east-west with an area of 1,700 m^2. The center piece of the Great Bath was the pool, a sunken rectangular basin approximately 12 × 7 m and 2.4 m deep. The sewage system was a network of effluent drains (built of brick masonry) constructed along the streets. These drains, located along one side of the street, were U-shaped approximately 50–60 cm deep. The drains were built of bricks set in clay mortar with covers made of loose bricks, flagstones or wooden boards. Covers could be removed for cleaning purposes. Wall drain chutes were used through which effluent flowed into the public drain or into a catchment basin. So in the 3rd millennium B.C. time period the Indus civilization had bathrooms in houses and sewers in streets and the Mesopotamians were not far behind (Adams, 1981).

1.3.2 Mesopotamia

Earliest city states of Mesopotamia appeared in the southern part of the Mesopotamian plain with the civilization known as the Sumer. Akkadian was a Semitic language used in this area starting in about 2600 B.C. The Akkadian word for this part of Mesopotamia was Sumer (Sumerian), the modern day word for this civilization. The Sumerian civilization included a group of cities that emerged around 3000 B.C.; however, ancestors of the Sumerian civilization were in Mesopotamia much earlier. The plains had few natural resources, limited rainfall, and no timber, stone, or metals. The earliest of the urban settlements, Eridu (see Fig. 1.4), was during the fifth millennium B.C. Eridu, centered on a temple complex built of mudbrick near the Euphrates River in a small depression that allowed water to accumulate. The location was at the location of a marsh, desert, and alluvial soil, and had a constant supply of water. Eridu was not really a city, but its neighbor Uruk (see Fig. 1.4) could be called a city. Between 3800 and 3200 B.C. was the most important period of Uruk's history. Uruk's growth was to cover an area of

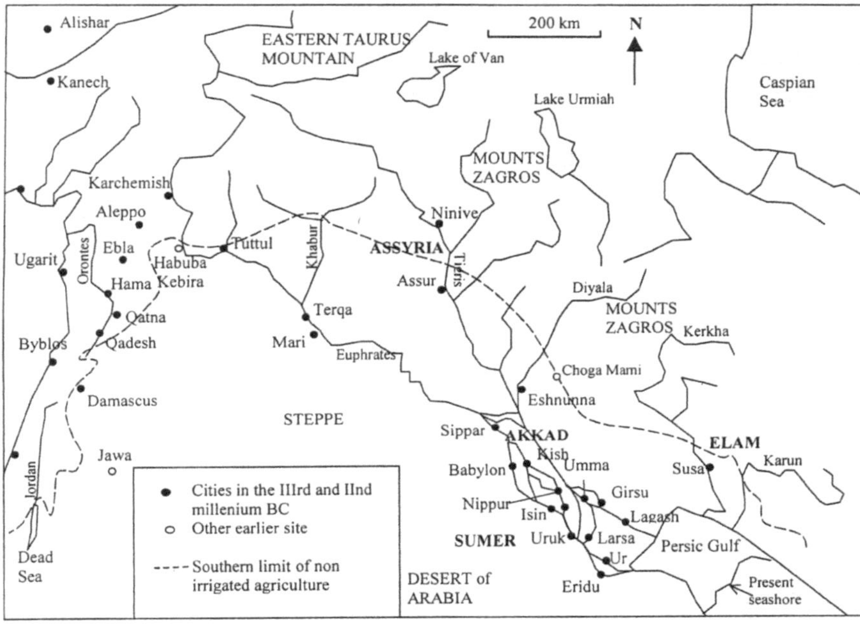

Fig. 1.4 Tigris and Euphrates Rivers with the major cities during antiquity (Violet, 2006)

about 550 ha (about half the size of Rome at its peak in A.D. 100). Uruk found ways to sustain its economic vigor over the centuries. However in around 3100 B.C. the trading links disappeared, possibly because the water supplies around Uruk began to dry up or the land was overly cultivated that the rural economy needed to support the city collapsed (Freeman, 2004). Uruk's demise was most likely connected to the rise of many smaller city states each developing its own access to water and surrounding land. Around 2330 B.C. southern Mesopotamia was conquered by Sargon of Akkade, history's first recorded emperor.

The sources of water for urban centers during early civilizations included canals connected to rivers, rainwater harvesting systems, wells, aqueducts, and underground cisterns. Figure 1.4 shows the major urban areas during the early Bronze Age. Table 1.1 summarizes the sources of water for urban centers of the Bronze Age (4000–1100 B.C.). In ancient Mesopotamia, during the Bronze Age urban centers of the Sumer (Sumerian) and Akkud (Akkadian) (III millenium B.C.) had a canal(s) connected to the Euphrates River or a major stream for both navigation and water supply for daily uses. In Mari a canal connected to the city from both ends and passed through the city (Viollet, 2006). Servant women filled the 25 m^3 cistern of the palace with water supplied by the canal. Later on other cisterns were built in Mari and connected to an extended rainfall collection system. Terracotta pipes were used in Habuba Kebira (in modern Syria), a Sumerian settlement in the middle Euphrates valley in the middle of the IV-th millennium B.C. (Viollet, 2006).

Table 1.1 Summary of sources of water for cities, settlements, and palaces of the early civilizations (4000–1100 B.C.)**

Source of water	Cities (Palaces)
Short canal connected to permanent river	Uruk, Ur, Babylon (all cities in the Tigris and Euphrates valleys)
Canals and reservoirs storing flood water of nonpermanent river, rainfall	Jawa, Khirbet el Umbashi
Rainwater harvesting (gutters and cisterns)	Agia Triadha, Chamaizi, Mari, Knossos, Myrtos-Pygros, Phaistos, Zakross
Wells	Ugarit (Syria), Palaikastro, Knossos, Zakros, Kommos, Mohenjo Daro (Indus Valley)
Aqueducts from source at altitude	Knossos*, Mallia*, Tylissos, Pylos, Thebes, DurUntawsh (Elam)
Underground cisterns w/ steps	Mycenae, Athens, Tyrins, Zakros, Tylissos
Springs	Knossos, Tylissos, Syme

*Probable.
** Information on Mesopotamian sites and Indus Valley is from Viollet (2006).

Inscriptions of Sennacherib (son of Sargon II) refer to a great network of canals, often describing them in the context of elaborate gardens and parks. Sennacherib was the king of Assyria (705–681 B.C.) succeeding his father. He relocated the imperial capital from his father's city of Khorsabad to the long established town of Nineveh, which he enlarged. For the new imperial capital he started the development of an elaborate water supply system. He constructed, in phases, a large canal network for not only agriculture, but also to sustain life in the new urban area, part of the Assyrian urbanism. Ur (2005) reassessed Sennacherib's canal network using aerial photography and satellite imagery. The project was most likely constructed in four phases: the Kisiri canal (year built, 702 B.C.); the Mount Musri canal (694 B.C.); the Northern System (ca. 690 B.C.); and the Khinisi system (ca. 690–688 B.C.). These years are based on work by Avrial M. Bagg (2000) as reported in Ur (2005). The canals received water from both rivers and springs and the canals followed the natural terrain (Ur, 2005). The canal lengths, as reported by Ur (2005) are 13.4 km for the Kisiri canal; 46.4 km for the Northern System, and 55 km for the Khinis canal. To give an idea of the canal sizes, the canal near Bandwai had been excavated 80 m wide and 20 m deep (Ur, 2005).

1.4 The Early Greeks

1.4.1 Minoans

The Minoan culture flourished during the Bronze Age in Crete. A systematic evolution of water management in ancient Greece began in Crete during the early Bronze Age, i.e. the Early Minoan period (ca. 3500–2150 B.C.). Wells, cisterns, water distribution, fountains, and even recreational functions existed. In prehistoric Crete rivers and springs provided people with water. Starting the Early Minoan period II (ca. 2900–2300 B.C.), a variety of technologies such as rainwater harvesting, wells, cisterns, gutters, channels, sedimentation tanks, and aqueducts were used. Also the Minoan architecture included flat rooftops, light wells, and open courts played an important role in the water management. The rooftops and open courts acted as catch basins to collect rainwater which flowed to storage areas or cisterns. Table 1.1 provides the sources of water supply for some of the Minaon palaces and settlements.

During the Neopalatial period, ca. 1700–1400 B.C., Knossos was at the height of its splendor. The city extended an area of 75,000–125,000 m^2 and had an estimated population in the order of tens of thousands of inhabitants. The water supply system at Knossos was most interesting; however, as Graham (1987) points out the sources and methods of supplying water are only partially understood. There were wells and an advanced system of rainfall collection for water supply. Terracotta pipe conduits (60–75 cm flanged to fit into one another and cemented at the joints) were used within the palace for rainfall collection and/or water distribution as shown in Fig. 1.5a. Possibly the piping system was pressurized. A water distribution system also makes possible the existence of an aqueduct. An aqueduct made of terracotta pipe could have crossed a bridge on a small stream south of the palace which carried water from a perennial spring on the Gypsadhes hill (Graham, p. 219, 1987). Figure 1.5b shows a drainage channel and Fig. 1.5c shows a rainfall harvesting system made of carved stone that collected water from the roof and directed it to a cistern. Unfortunately, around 1450 B.C. the Minoan palace was destroyed.

A very interesting drainage system also exists on the northeastern side of the palace as shown in Fig. 1.5d. Alongside the stairway is a small channel consisting of a series parabolic-shaped stepped chutes that convey rainwater down steam to the sedimentation tank or basin. According to Viollet (2003) each step has a vertical drop of 16 cm and the length of each step is 43 cm with the vertical distance from the top of the step down to the top of each chute is 13 cm.

The Minoan settlements used rainfall collection and cisterns over a 1,000 years before the classical and Helenistic-Greek cities. Cisterns were used to supply water (store runoff from roof tops and court yards) for the households through the dry summers of the Mediterranean. The two earliest large cisterns of Minoan Crete were built in the first half of the IInd millennium B.C., which was the time of the first Minoan palaces at Myrtos-Pyrgos (Cadogan, 2006). These cisterns remain an unusual attribute of the Minoan settlement, as the Myrtos River has been able to supply water to the base of the Pyrgos hill. Both cisterns are circular with vertical walls and rounded bottom. The walls and bottom are coated with white lime plaster 1–2 cm thick (Cadogan, 2006). Similar round structures exist at Knossos, Mallia, and Phaistos, which have been called granaries, but according to Cadogan improbable because of the locations at the bottom of hills. It would have been difficult to prevent water from running into the round structures during a storm (as illustrated in Fig. 1.6a). The main drain at the southern end of the palace is shown in Fig. 1.6b.

The hill of Phaistos was settled first at the end of the Neoloithic period (4th million B.C.). Later in the early Bronze age (during the Early Minoan period) the

(a) (b) (c) (d)

Fig. 1.5 Water components in Knossos. (**a**) Pipes made of terracotta. (**b**) Drainage channel. (**c**) Carved stone elements of rainfall harvesting system collecting water falling from roof. (**d**) Stepped water channel and sedimentation (desilting basin). Along the stairway is a small channel (for rainwater collection) consisting of a series parabolic-shaped stepped chutes that convey rainwater down stream to the sedimentation tank or basin (copyright permission with L.W. Mays). Color version available in Appendix

Minoans built above the ruins of the Neolithic houses. At the end of the Prepalatial period Phaistos became very prosperous with the construction of the first palace c. 2000/1900 B.C. A large part of the water supply system in the Palace of Phaistos, as well as in other cities and villages in Minoan Crete depended directly on rainfall, collected from roofs and courtyards and then directed to cisterns. The drainage

(a) (b)

Fig. 1.6 Water components in Phaistos (**a**) Courtyard used for rainfall harvesting with cisterns (*round structures*) shown in background to the right. (b) Exit of main drain at southern end of palace (copyright permission with L.W. Mays). Color version available in Appendix

systems were in some cases conveyed into terracotta vessels near light wells, which acted as water collectors.

Tylissos was one of the important cities in Ancient Crete during the Minoan era, flourishing (2000–1100 B.C.) as a peripheral center dependent on Knossos. The water supply system to Tylissos included an aqueduct from the spring of Agios Mamas (Sklivaniotis and Angelakis, 2006). The portion of the aqueduct shown in Fig. 1.7(a) is located at the entrance of the three villas of Tylissos. Infiltration devices made of terracotta and filled with charcoal (burnt wood) were found at the spring. They are now located at the Archaeological Museum in Heraklion. This infiltration device essentially acted as an activated carbon process. The water distribution system was constructed of closed pipes and curved stone channels. The water supply system to Tylissos included an aqueduct developed in the Minoan period that was constructed of closed pipes and curved stone channels (Fig. 1.7a). Secondary conduits were used to convey water to a sediment tank constructed of stone (see Fig. 1.7b and c), used to remove sediment and/or suspended sediments. Also note the hole on the lower part of the tank shown in Fig. 1.7c used to drain the tank for cleaning. Water then flowed from the sedimentation tank through the small channel shown in Fig. 1.7d to the main cistern for water storage. Steps, shown in Fig. 1.7e were used to descend down to the various water levels.

1.4.2 The Mycenaeans

The Mycenaean civilization (named from the site of Mycenae in northeastern Argolis in Peloponnese of southern Greece) flourished between 1600 B.C. and ca. 1100 B.C. when it perished with the decline (collapse) of the Bronze-Age civilization. The Mycenaeans built several dams that were essentially long, low dikes in Peloponnese and Beotia as reported by Knauss (1991a,b) with the dimensions given.

Fig. 1.7 Water system in Tylissos, Crete. (**a**) Aqueduct bringing water from springs. (**b**) Sedimentation tank in foreground with stone channel connecting to cistern. (**c**) Sediment tank (**d**) Channel connecting sediment tank to cistern. (**e**) Steps leading down to cistern (photos copyright by L.W. Mays). Color version available in Appendix

In Peloponnese these included: Pheneos (2.5 m high, 2,500 m long); Stymphalos (2.5 m high, 1,900 m long); Orchomenos (2 m high, 2,100 m long); Mantinea (3 m high, 300 m long); and Taka (2 m high, 900 m long). In Beotia these included: Boedria (2 m high, 1,250 m long); Thisbe I (2.5 m high, 1,200 m long); and Thisbe II (4 m high, 200 m long). The Tiryns Dam (located four km east of Tiryns near the modern village of Ayios Adrianos) and a 1.5 km long diversion canal were built to divert flood flows from entering the town in the small river that flowed through it to the sea (Zangger, 1994). This diversion canal diverted the flood flows to another river that also flowed to the sea. Tiryns Dam was 10 m high and 100 m long constructed with an earth core and two masonry walls. Width of the dam was 103 m on the right bank and 57 m on the left bank.

The Mycenaeans also built other water projects such as cisterns. Taylour (1983) discussed an underground cistern used to store water from the Perseia Spring in the XIIIth century B.C. A 200 m long underground conduit was excavated into rock and terracotta pipe was used. A 2 km long aqueduct partly made of wood, dug channel in rock and U-shaped terracotta sections was used to bring water to the Palace of Pylos in southwestern Peloponnese (Taylour, 1983).

1.5 Greeks

1.5.1 Archaic and Classical Periods

In the archaic (750–480 B.C.) and the classical (480–323 B.C.) periods of the Greek civilization, aqueducts, cisterns, and wells were similar to those built by the Minoans and Mycenaeans. However, the scientific and engineering progress during those stages enabled the construction of more sophisticated structures. One of the most famous is the tunnel of Eupalinos (530 B.C.) on Samos Island, the first deep tunnel in history that was dug from two openings with the two lines of construction meeting near the middle. The construction of this tunnel, which served the water supply of Samos, was made possible by the progress in geometry and geodesy that was necessary to implement two independent lines of construction that would meet (Koutsoyiannis et al., 2007).

There are several other known aqueducts in Greek cities as water supply was regarded as an essential and necessary infrastructure of any city. For safety reasons, aqueducts were always subterranean, either tunnels or trenches. At the entrance of the city, aqueducts would branch in the city and would feed cisterns and public fountains in central locations. Along in the bottom of trenches or tunnels of aqueducts, pipes usually made from terracotta were laid, allowing for protection. One, two or more pipes in parallel were used depending upon the flow to be conveyed. The terracotta pipe segments (20–25 cm in diameter) fit into each other and allow access for cleaning and maintenance by elliptical openings (Fig. 1.8) that were covered by terracotta covers. This is one indication of the awareness that the Greeks had for hygienic conditions (Koutsoyiannis et al., 2007).

The Peisistratean aqueduct, constructed in Athens during the time of the tyrant Peisistratos and descendents ca. 510 B.C. This aqueduct carried water from the foothill of the Hymettos mountain (probably from east of the present Holargos suburb) for a distance of 7.5 km to the center of the city near the Acropolis (Tasios, 2002). Figure 1.8 illustrates the pipe segments. Many wells and later cisterns were needed for the water supply system. The greater part of the aqueduct was carved as a tunnel at depths reaching 14 m. Other parts of the aqueduct were constructed as a

Fig. 1.8 Peisistratean aqueduct (**a**) Terracotta pipe segments laid in a channel. (**b**) Lead pipe joint and elliptical pipe opening for cleaning joint. Color version available in Appendix. (**c**) and (**d**) Rectangular shaped conveyance channels with cover. (**e**) Pipe segment showing cover for pipe opening (photos copyright by L.W. Mays)

channel carved in rock or made from stone masonry, with depths of 1.3–1.5 m and a width of 0.65 m (Papademos, 1975). The terracotta pipe segments were placed in the bottom of the tunnel or channel (Fig. 1.8a). Figure 1.8b shows the elliptical pipe opening which was used for cleaning purposes. Figure 1.8c and d show a rectangular shaped channel and Fig. 1.8e shows a pipe segment with cover for pipe opening.

Another great achievement during the Archaic Period was the tunnel of Eupalinos (or Eupalinos aqueduct, named after the builder who was from Megara) on the island of Samos, built in the 6th century B.C. The tunnel was excavated through a mountain (Mt. Kastron) from both ends, having a length of 1,036 m. It is located 55 m above sea level and 180 m below the top of the mountain. Construction started in 550 B.C. and lasted for 10 years during the tyranny of Polycrates. For further reading refer to Apostol (2004), Burns (1971) and Van der Waerden (1968).

As previously addressed the first evidence of the use of cisterns in Greece was during the Minoan era (Antoniou et al., 2006). One example of a cistern is located in the city-state of Dreros on Crete during the Classical Greek period. Dreros was built on a saddle between two peaks, on the slope of mount Kadistos. This city had an agora (market place) approximately 30 × 40 m^2 (Antoniou et al., 2006) with the cistern located at the uphill side of the agora as shown in Fig. 1.9a. This cistern is shown in Fig. 1.9b, illustrating the steps down into the cistern. Myers et al. (1992) reported that this cistern is the first and largest cistern ever known. The rectangular shape of the cistern has dimensions of $13 \times 5.5 \times 6.0$ m^3 (Antoniou et al., 2006).

(a) (b)

Fig. 1.9 Location and remains of the central cistern in the agora of Dreros (a) Agora looking toward cistern. (b) Steps leading down into cistern. (copyright permission with L.W. Mays)

1.5.2 Hellenistic Period

Later, during the Hellenistic period [323 B.C. (death of Alexander the Great)–146 B.C. (conquest of Greece by the Roman Empire)], further developments were made by Greeks in hydraulics, such as in the construction and operation of aqueducts, cisterns, wells, harbors, water supply systems, baths, toilets, and sewerage and drainage systems. During that period the political and economic situation changed leading to much more architectural development and urban beautification, of which aqueducts played a major role. During the beginning of the Hellenistic era several cities like Athens, Samos and Olynthos already had aqueducts. Probably during the middle of the 2nd half or the 3rd century B.C. the first aqueduct was constructed in Pergamon (Pergumum) bringing water from a spring in the mountains north of the city. This aqueduct was around 15 km and was constructed with fired clay pipes having an inner diameter of 13 cm and lengths up to 60 cm (Fahlbusch, 2006). Aqueduct pipes were laid in an excavated bed a little below the natural soil surface. Various stamps were imprinted on the pipes. This pipeline was constructed during the reign of King Attalos I and therefore has been called the Attalos aqueduct. An interesting note is that there is a saddle north in the mountain which the aqueduct was constructed so that the aqueduct outlet was at an elevation of 25 m higher than the crest of the saddle so that a pressurized pipeline, inverted siphon, was necessary. This very well could have been the first large-scale application of pressurized pipes during antiquity (Fahlbusch, 2006).

The progress in science during the Hellenistic period provided a new technical expertise. Hellenistic aqueducts usually used pipes, as compared to the Roman masonry conduit. Furthermore, following the classical Greek tradition, the aqueducts continued to be subterranean for security reasons (not to be exposed to aliens, e.g. in case of war) but also for the safety of the construction during earthquakes which are frequent in the area. This again contrasts the Hellenistic technology with the later Roman technology, whose apparent characteristic was the use of the arches and aqueduct bridges.

Greek aqueducts generally operated by free surface flow. However, during the Hellenistic period, the scientific progress in understanding hydrostatics and water and air pressure (due to Archimedes, Hero of Alexandria and others; Koutsoyiannis et al., 2007) allowed the construction of inverted siphons at large scales (lengths of kilometers, hydraulic heads of hundreds of meters). Hellenistic engineers constructed inverted siphons to convey water across valleys in aqueducts of several cities including Ephesus, Methymna, Magnesia, Philadelphia, both Antiochias, Blaundros, Patara (see Fig. 1.10), Smyrna, Prymnessos, Tralleis, Trapezopolis, Apameia, Akmonia, Laodikeia and Pergamon. These siphons initially were constructed of terracotta or stone pipes (square stone blocks to which a hole was carved as shown in Fig. 1.10). The need for higher pressures led to the use of metal pipes, specifically made from lead. Thus, one of the aqueducts of Pergamon includes an inverted siphon, constructed of lead pipes, over 3 km in length with a maximum pressure head of about 180 m.

The technology of cisterns developed through history and possibly peaked during the Hellenistic period. Because of the lack of adequate water supplies on most Greek islands water cisterns were used. During the Hellenistic period the technology of cisterns showed progress. During this time the water supply in several cities all over Greece was dependent entirely on precipitation. Rainwater was collected from the roofs, yards, and other open spaces (Angelakis and Spyridakis, 1996; Antoniou et al., 2006).

Aristotle in his *Politics* (vii, 1330b) written in the 320's B.C. asserts that "cities need cisterns for safety in war." During this time a severe 25-year drought required the collection and saving of rainwater (Camp, 1979). Also about this time cisterns were built in the Athenian Agora for the first time in centuries (Parsons, 1943; Crouch, 1993).

Fig. 1.10 Inverted siphon at Patara. Color version available in Appendix (copyright permission with L.M. Mays)

1.5.3 Greek Water Management

"Since the area is sufficiently with water, either from continuous flow rivers, or lakes or rich springs, but most people used artificial wells, Solon (594 B.C.) made a law that if there was a public well within a hippicon, that is, four stadia (4 furlongs, 710 m), all should use that; but when it was further off, they should try and procure water of their own; and if they had dug ten fathoms (18.3 m) deep and could find no water, they had liberty to fetch a hydria (pitcher) of six choae (20 l) twice a day from their neighbors; for he thought it prudent to make provision against need, but not to supply laziness." (as described later by Plutarch (47–127 A.D.))

1.5.4 Greek Hydraulic Technology of Devices

The late-classical/Helenistic city of Priene on the Anatolian west coast of present day Turkey had a main drainage canal with an unusual masonry outlet structure that allowed self-cleaning of the drainage outlet. This outlet structure contained a doubly curvilinear, contracting rectangular cross-section flow passageway that allowed drainage water to flow through the perimeter wall of the city. Hydraulic analysis by Ortloff and Crouch (1998) showed that the internal shape of the structure causes the flow to create multiple circulatory mixing flows that agitated and entrained debris in the outflow stream, self-cleaning the outlet and preventing clogging. Once again we see that the Greeks possessed a high level of awareness about the hygienic conditions needed for health. Priene must have been well planned because of the placement of the underground water supply network supplied from a system of reservoirs. In addition there was an elaborate channel drainage system to convey stormwater and waste out of the city.

Archimedes (287–212 B.C.), who has been considered by many as the greatest mathematician during antiquity, was the founder of hydrostatics and introduced the principle of buoyancy. He lived in Syracusa, located on Sicily. The foundation of hydraulics after Archimedes led to the invention of several hydraulic devices with applications ranging from the lifting of water to musical instruments. Archimedes's helix or water screw (see Fig. 1.11) is the first device characterized as a pump by modern standards. The invention of the water screw is linked to the study of the spiral, on which Archimedes wrote the treatise, On Spirals in 225 B.C. The water screw consists of a cylinder with a continuous screw that extends the length of the cylinder forming a spiral chamber. To operate the device the lower end is placed in the water and the screw is turned with the handle (Fig. 1.10a) raising the water to the higher elevation. The hydrostatic principles that Archimedes developed have proven to be the most enduring achievements of Greek mechanics.

The force pump, a water lifting device, was invented by Ctesibius of Alexandria (ca. 285–222 B.C.), who also invented other instruments like the water clock (clepsydra) and the *hydraulis* which is a water organ. Vitruvius, the Roman author of *De Architectura*, was the only ancient author to attribute the invention of the force pump to Ctesibius of Alexandria by calling the device – *Ctesibica machina* (x,7,1) and by

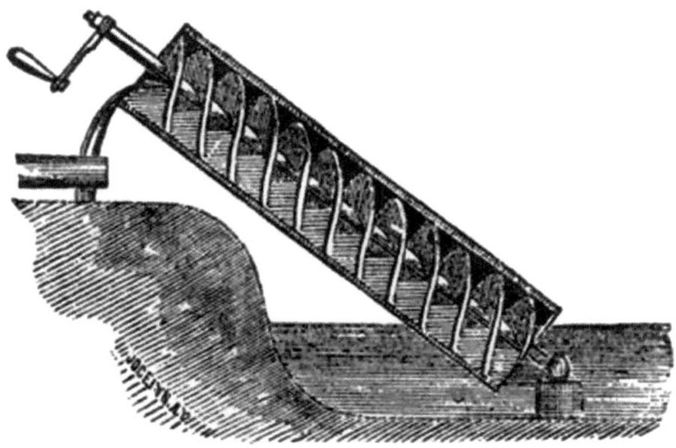

Fig. 1.11 Illustrations of the Archimedes Screw (a) Chambers Encyclopedia, J.B. Lippincott Company, Philadelphia, 1875

the comment in (x,7,4) *nec tamen haec sola ratio Ctesibii fertur exquisita* (and this is not the only choice device of Ctesibius current) (Oleson, 1984). The force pump is described in a later chapter on Romans in a section of Roman hydraulic devices. The principal of the siphon has been attributed to him. Further discussion of the force pump is given in Chapter 7.

1.5.5 Water Clocks

Simple water clocks date back to perhaps 1500 B.C. in Egypt and Mesopotamia. The Greeks called the water clock a clepsydra (water thief). One example of the water clock or clepsydra is Ctesibius's in the 3rd century B.C. illustrated in Fig. 1.12. This water clock had a constant-head inflow from a water supply tank (above the drum with the wedges). Note how water flowed into the partitions or wedges of the drum from the constant head above. Ctesibius probably tried to regulate the outflow from the supply tank by using a valve adjusted by the wedges. As the wedge in the drum receives water from above, the drum rotates causing the gear mechanism to rotate. Rotation of the series of gears causes the cylinder (with the scale) to rotate indicating time. Figure 1.13 shows the Tower of the Winds, an octagonal clock tower, below the Acropolis in Athens, which housed a water clock. The 12 m tall clock tower was built in the first half of the 1st century B.C.

1.6 Water Supply at the Urartian Capital of Tuspa

The name Urartu first appeared in Assyrain inscriptions around 1250 B.C. Harrian tribes living in the present day Armenian highlands near lakes Van, Urmia, and Sevan (to the north), were harassed by the Assyrian campaigns. Lakes Van and

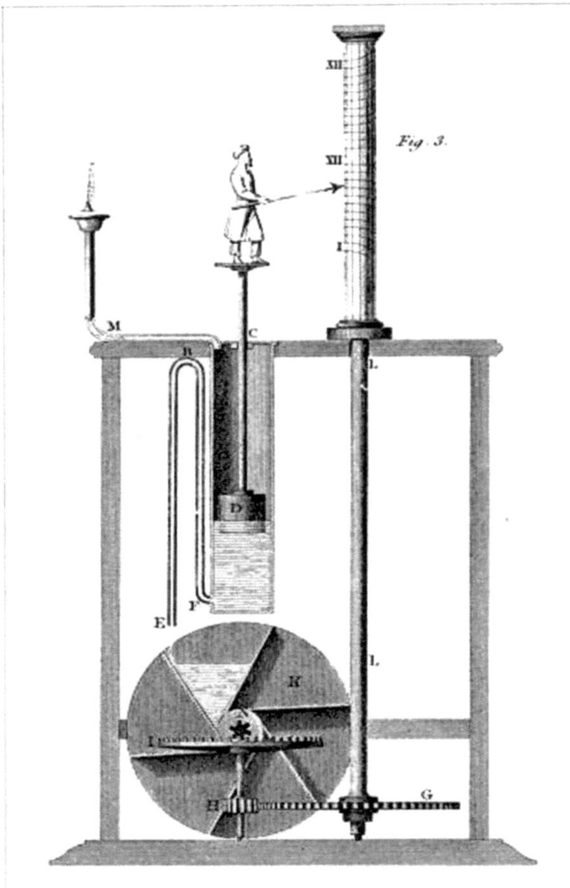

Fig. 1.12 Illustration of Ctesibius's water clock (clepsydra). Abraham Rees (1819) Clepsydra, Cyclopædia: or, a New Universal Dictionary of Arts and Sciences. Probably illustrated by John Farey, Jr. (1791–1851). (Figure in public domain) http://www.antique-horology.org/_editorial/clepsydra/

Urmia are shown in Fig. 1.4. These tribes united to form the kingdom of Urartu around 850 B.C. and between 850 and 600 B.C. Today this region of the kingdom of Urartu is divided among Armenia, eastern Turkey, and northwestern Iran. Urartians constantly fought battles with Assyria. The Urartians had considerable artistic and technical skills, especially water technology building significant irrigation works and water supply systems. Urartians were the only people in the Near East to have the elaborate and well-planned water supply systems comparable to those in Egypt and Mesopotamia (Garbrecht, 1980).

Lake Van has no apparent outlets and the water is brackish so that it could not have been used for water supply or irrigation. Urartians first exploited the spring in the valley of Engil Cayi, shown in Fig. 1.14. Water from the spring flows naturally to the Engil Cayi; however, it was collected immediately below the spring using a simple stone and earthworks, then channeled across the Engil Calyi to its right shore in an aqueduct bridge. Even today the water is collected below the spring

Fig. 1.13 Tower of the Winds. Located below the Acropolis in Athens, designed by the famous astronomer Andronikos of Kyrrhos to be an elaborate water clock (on the inside). The name is derived from the personifications of the eight winds carved on the eight sides of the building. Originally thought to have been built in the 1st century B.C. during the early Roman Empire, but now many archaeologists think the construction was in the mid-2nd century B.C. during the Hellenistic period (copyright permission with L.W. Mays). Color version available in Appendix

and a concrete structure is used most likely similar to the Urartian times. Along the ancient canal supplying water to Tuspa there are 14 inscriptions that confirm that it was constructed by King Menua, who reigned between 805 and 785 B.C. The route of the Menua canal is shown in Fig. 1.14. Menua canal is about 56 km long to the plain of Van-Kale where the water irrigated the fields for the ancient city of Tuspa, as it does today for the city of Van. In 1956 the middle part of the canal had to be replaced by a modern concrete channel because the maintenance of the old one became to expensive (Garbrecht, 1980). Essentially the canal has flowed uninterruptedly for 2,500 years.

The capital of the Urartian kingdom was moved to Toprak-Kale (ancient Rusahinili or Rusa city) around 700 B.C. after which the Menua canal was still used to bring water to Tuspa. The new capital needed a water supply which was developed by transforming the flat water basin of Kesis Goluwas, located about 30 km away and at a 900m higher elevation, into a lake, Lake Rusa). The lake was named after either Rusa I who reigned 730–714 B.C. or Rusa II who reigned about 685–645 B.C. The basin was dammed at its natural outlets, dams 1 and 2 in Fig. 1.14.

Flow from the north dam (1) flowed through the valley of Engusner Cayi directly to Toprak-Kale (Rusahinili). Dam 3 (called Faruk Bendi today), downstream of dam

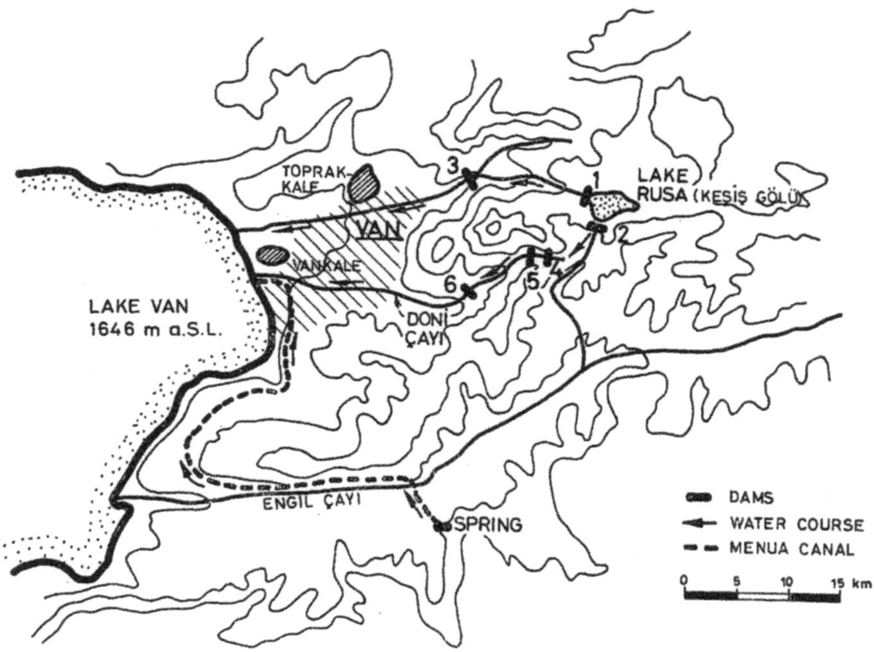

Fig. 1.14 Water supply of ancient Urartian capital of Tuspa (Van) (Garbrecht, 1980)

1 half way between the reservoir and the city, may or may not have been built by the Urartians. The north dam was reconstructed in 1894–1895 and again in 1952. The outlet is capable of discharging 2.5–3 m^3/s (Garbrecht, 1980).

Flow from the south dam (2) would reach Lake Van through the valley of Engil Cayi. To use this water for the plain, the passage through the south wall was redirected to the drainage area Doni Cayi. Figure 1.14 shows three of the reservoirs (dams 4, 5 and 6) in the valley of Doni that are still evident of the ten reservoirs that were built (probably not all were built by the Urartians). Reservoirs created by dams 4 and 5 are still in use today for storing irrigation water for the Van plain (Garbrecht, 1980).

This is certainly one of the ancient mega hydraulic projects that resulted from thoughtful planning and excellent workmanship evidenced by its 2,500 years of operation, still serving part of its original purpose.

1.7 Nabataean City of Petra's Water Supply

The extent of the Nabataean kingdom (flourished 168 B.C.–A.D. 106) is not known with certainty, but some historian that think during the peak it stretched from modern day Yemen to Damascus and from western Iraq into the Sinai Desert (Ortloff, 2005). Others define the kingdom as including Jordan, the Hawran in southern Syria, the

Sinai, the Negev, a large part of the Hijaz in north-western Arabia and for a short time Damascus (Ruben, 2003). The empire was positioned between the Egyptian, Babylonian, and Assyrian empires so that there were many influences that dominated the Nabataean culture. The use of pipelines, canals, and cisterns, and even qanats was established long before Nabataean times. These previously established technologies were surely available during the planning and development of the Petra's water system. The ancient city of Petra is located in southwestern Jordan, situated in a large open valley surrounded by ragged mountains with limestone slopes at the top and sandstone down to the wadis. Figure 1.15 shows a map of the ancient city of Petra.

A constant year-round water supply for the urban population was made difficult because of the variation in seasonal rainfall and spring flows. Limited water resources from rainfall and springs along with the mountainous terrain required unique thinking by the Nabataeans to apply water technologies. The Nabataeans built dams, diversion walls, and terraces, and dug cisterns and reservoirs to store water. They built channels and aqueducts to bring water from springs. The verb *nabat* in Arabic means for water "to percolate from underground to the surface," derived from *anbata* meaning "to dig for water" or "to draw water from underground."

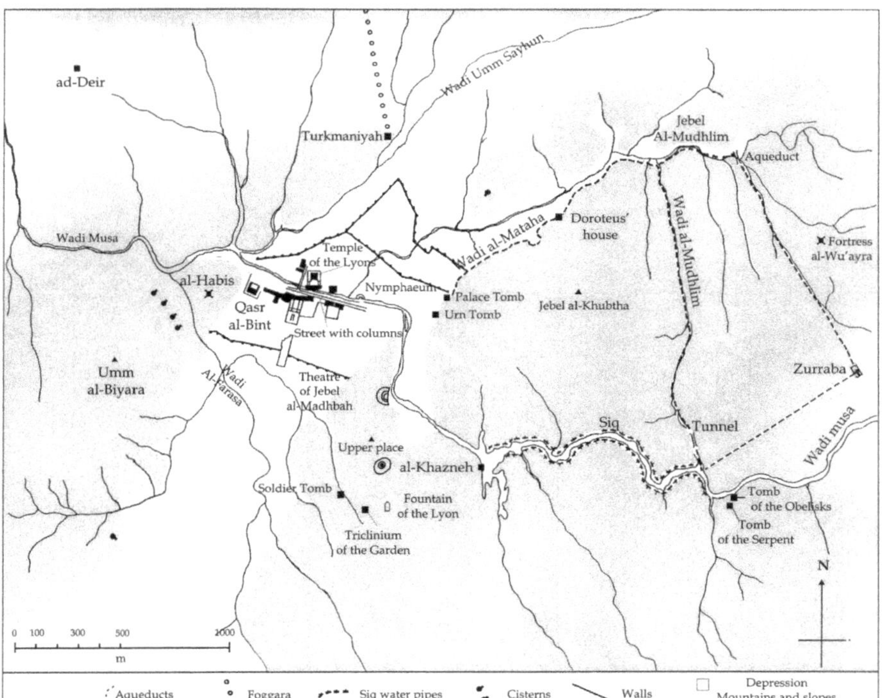

Fig. 1.15 Map of Petra (Laureano, 2001)

1.7.1 The Siq

The entrance to Petra is now called the Siq, which had been the natural continuation of Wadi Musa before the Nabataeans diverted it to the Wadi Mudhlim. A dam was first constructed across the mouth of the Siq during the 1st century B.C. (Oleson, 1995) in order to protect the central area of Petra. This dam prevented water from flowing through the Siq and was rerouted through an 82-m long tunnel (Figs. 1.16 and 1.17) excavated through rock leading to the Siq al-Mudhlim then through the Wadi Mudhlim and Wadi Mataha, rejoining the Wadi Musa. Figure 1.16 shows the reconstruction of the situation at the entrance to the Siq and Fig. 1.17 shows the tunnel excavated by Nabataeans to divert flood water at the entrance of the Siq from

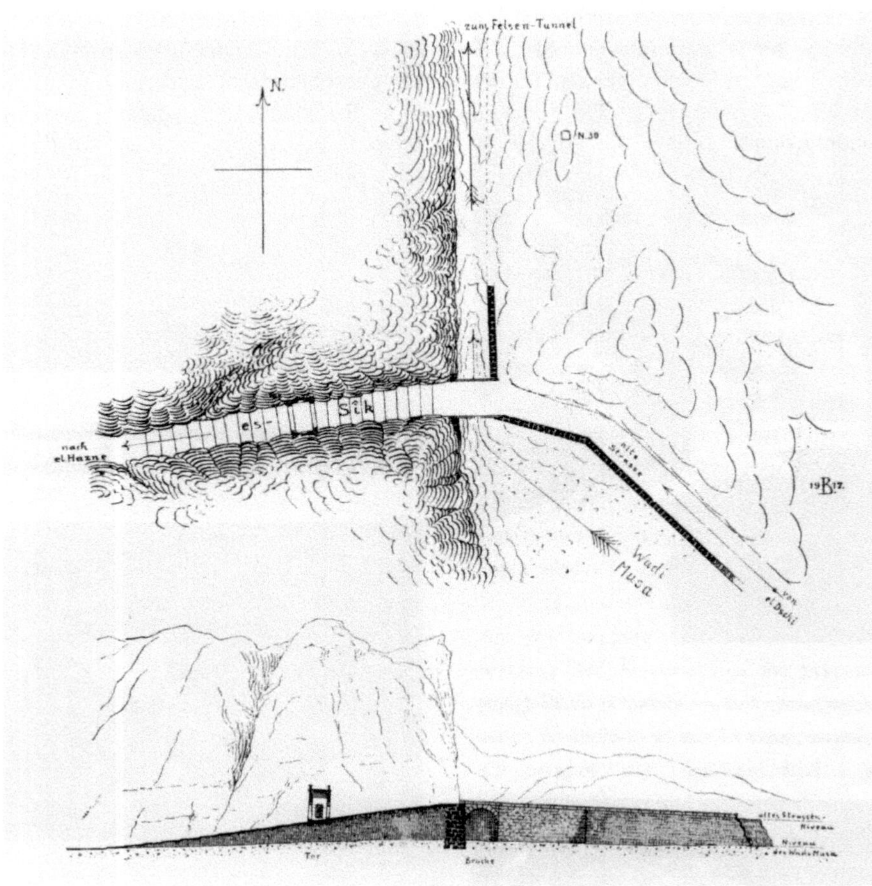

Fig. 1.16 W. Bachmann's 1917 reconstruction of the situation at the entrance to the Siq, with the plan above and the elevation below. (Bachmann et al., 1921)

Fig. 1.17 Tunnel excavated by Nabataeans to divert flood water at the entrance of the Siq from Wadi Musa into Siq Mudhlim tunnel (photo courtesy of Mac McKee)

Table 1.2 Summary of the sequence of major events of the Siq (after Ruben, 2003)

Activity	Date
First trampled road	Beginning of settlement until early 1st century B.C.
Improved first road and first gravity flow channel on north side of Siq (right bank going downstream)	Sometime in first half of the 1st century B.C.
Paved street, dams in wadi inlets	Third quarter of 1st century B.C.
Clay pipeline on north side	Mostly completed by 30–25 B.C. and completely finished by 25–0 B.C.
Paved street and all its installations finished	In use by very end of 1st century B.C.
Southern gravity flow channel	Between 20–70 A.D.
Earthquake destruction	363 A.D.
Post earthquake repairs	Late 4th century A.D.

Wadi Musa into Siq Mudhlim tunnel. Table 1.2 is a summary of the sequence of major events of the Siq (Ruben, 2003):

Four watershed catchment areas contribute water to the Siq, which are Khubtha, Jilf, Qantara, and Madrass, with a total area of just over a 1 km^2. In these four catchment areas there were a large number of hydraulic features built by the Nabataeans, including 136 terrace barriers, 143 wadi barriers, 26 dams, and 7 cisterns (Ruben, 2003). Oleson (1995) discusses the origins and design of the Nabataean water-supply systems.

(a) (b)

Fig. 1.18 The Siq (**a**) The gravine with the water channel (elevated bench flume) along the side. (**b**) Piping elements in the Siq showing clay pipes with bell and spigot joints filled with mortar (photos courtesy of Mac McKee)

Within the Siq (Fig. 1.18a) a spring water supply system consisting of a pipeline and a gravity flow channel that brought water from Ain Musa to the city. A pipeline (see Fig. 1.18b) was built on the north side using clay pipes which would have had pressurized flow (with a diameter of 18 cm). The channel on the southern side was 34 cm wide, 18 cm deep, and a slope of 4.9 percent with an average level of 1.5m above the floor of the Siq (Ruben, 2003). This double water supply system became a model for other parts of the water supply system for Petra. The Ain Brak and Ain Debdebeh aqueducts were both built with both a clay pipeline (pressurized flow) and a gravity flow channel. Two other spring water supply lines to Petra, the northern Khubtha and the Ain Abu Ullayqa channels, were gravity flow channels.

Settling tanks were used in the southern gravity flow channels to collect precipitated deposits of lime and sediments from the water originating at the Ain Masa spring and the Reservoir Surraba. Seven settling basins were constructed along the gravity flow channel, on the southern side of the Siq, between the entrance dam and the exit to the Khazneh. These rock-cut settling tanks were plaster coated with dimensions of approximately 60 cm square and about 60 cm deep with rounded corners (Ruben, 2003). Distances between the tanks varied by large distances, not presently understood.

Drinking basins were constructed along the southern water channel, having the same shape and dimensions as the settling tanks. These basins were cut into the rock on the side of the Siq floor so that they were not built on the longitudinal axis of the channel. The basin was connected to the channel by a ceramic pipe connected to the bottom of the water channel and the basin.

1.7.2 Water Sources

Rainwater harvesting was used extensively in Petra, as it is the most archaic device that humans have used. In Petra this meant using the technology on the high places and on bare sandstone walls with many innovations. The Nabataeans used various types of cisterns that they carved out of the rock and waterproofed using chalk. These cisterns ranged from small pools on the highlands to catch runoff to rectangular shaped cisterns at the bottom of the natural drips. Small pools were carved out of the highland and evolved into bell-shaped cisterns. These large cisterns are similar to large rooms carved out of vertical walls into which complex canals and pipe networks flow (Laureano, 2001). What is so unique is that the Nabataeans used every slope and surface as a means to harvest rainfall and stored every water source from a few drops to the large floods. This is why Strabo, the 1st century geographer, described Petra as ornamented with fountains and basins (Laureano, 2001).

Khottara make use of traces of moisture and night condensation of fog and dew by harvesting the exudation of condensation (humidity) by dripping into tanks, cisterns, or channels that catch the water on the walls and conveys it to pools. These structures provide water all year round. On the other extreme, one cistern called

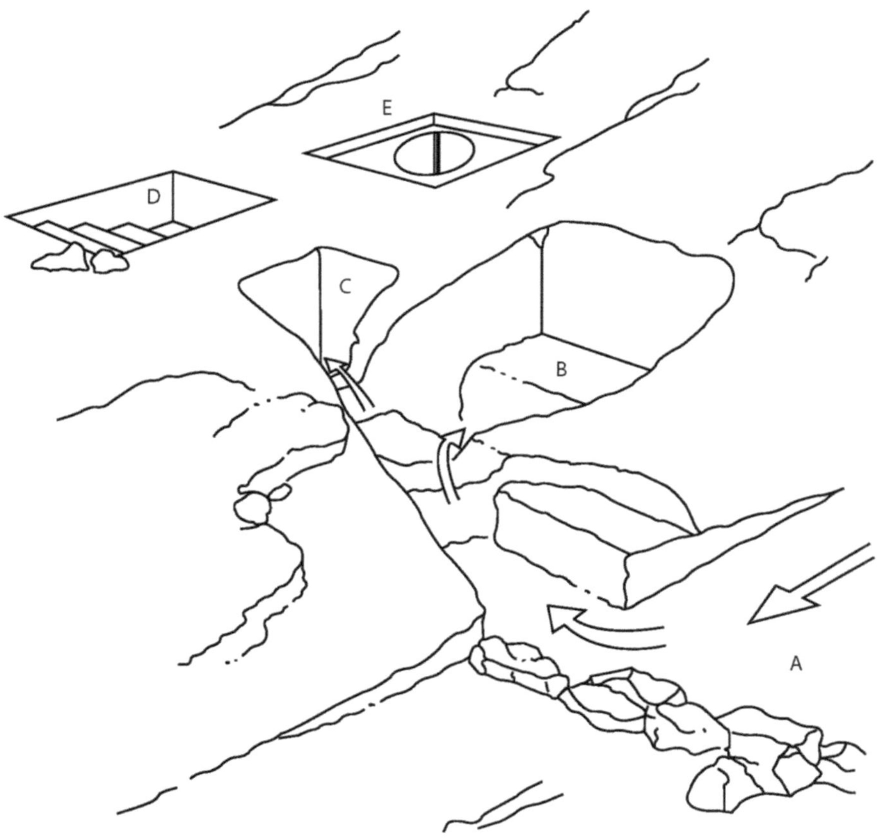

Fig. 1.19 Flood trap at Bir Huweimel (Petra). Flood water is diverted (**A**) and then cleaned by means of spillways in consecutive basins (**B**) and fills up the large underground cistern (**C**). A staircase (**D**) leads to the cistern and water is drawn from the cistern through the well (**E**) (Laureano, 2001)

Bir Huweimel at the bottom of Ras as-Slimane, actually traps flood water using a large room (depth of 9 m) excavated in the riverbed. Flood water is diverted to water intakes and decanting basins as shown in Fig. 1.19, to fill the large cistern where the water is stored. A staircase is used to enter the cistern and water is drawn from the cistern through a well.

The Ain Masa spring, approximately 7.0 km east of Petra was one of the water supply sources, which is very hard containing a high percentage of dissolved lime causing deposits in the pipeline and channel. Water also entered the Siq water lines from the Reservoir Surraba, at a height of over a 1,000 m just outside of Petra. Two aqueducts were constructed from the reservoir; one toward the south fed pipes of the Siq and the other to the north to the Wadi Al-Mataha and then along the Wadi to Petra.

1.8 Conclusions

In this chapter the development of water technology has been traced, beginning with its infancy in the development of irrigated agriculture and water supply for urban centers. We have looked at some of the major advances that occurred in hydraulic technology during antiquity, long before the development of the laws of conservation used in modern day design of hydraulic structures. This chapter does not discuss advances during the Roman period, which is covered in Chapter 7.

References

Adams R (1981) Heartland of cities: surveys of ancient settlements and land use on the central floodplain of the Euphrates. University of Chicago Press, Chicago, USA

Angelakis AN, Spyridakis SV (1996) The status of water resources in Minoan times: a preliminary study. In: Angelakis AN, Issar AS (eds) Diachronic climate impacts on water resources with emphasis on Mediterranean region. Springer-Verlag, Heidelberg, Germany, pp. 161–191

Antoniou G, Xarchakou R, Angelakis AN (2006) Water cistern systems in Greece from Minaoan to Helenistic period. In: Angelakis AD, Koutsoyiannis D (eds) Proceedings 1st IWA International Symposium Water and Wastewater Technologies in Ancient Civilizations, National Agricultural Research Foundation, Iraklio, Greece, held 28–30 October 2006, pp. 457–462

Apostol T (2004) The tunnel of Samos, engineering and science, No. 1, pp. 30–40

Bachmann W, Watzinger C, Wiegand TH (1921) Petra. Wissenschaftliche Veroffentlichungen des Deutsch-Turkischen Denkmalschutzkommandos, Heft 3, Berlin

Burns A (1971) The tunnel of Eupolinus and the tunnel problem of Hero of Alexandria. ISIS 62(2):172–185

Butler SS (1960) Irrigation systems of the Tigris and Euphrates valleys. J Irrigation Drainage Division. ASCE 86(4):59–79

Butzer KW (1976) Early hydraulic civilization in Egypt: a study in cultural ecology, The University of Chicago Press, Chicago

Cadogan G (2006) Water management in Minoan Crete, Greece: the two cisterns of one middle bronze age settlement. In: Angelakis AD, Koutsoyiannis D (eds) Proceedings 1st

IWA International Symposium Water and Wastewater Technologies in Ancient Civilizations, National Agricultural Research Foundation, Iraklio, Greece, held 28–30 October 2006, pp. 447–56.

Camp JM II (1979) A drought in the late eighth century B.C. Hesperia 48: 397–411

Crouch DP (1993) Water management in ancient Greek cities. Oxford University Press, New York, USA

Fahlbusch H (2006) Water engineering and management in the classic civilizations, Proceedings, La Ingenieria Y La Gestion Del Agua a Traves de Los Tiempos, held at the Universidad de Alicante, Spain, with the Universidad Politechnica de Valencia, Spain

Freeman C (2004) Egypt, Greece, and Rome: Civilizations of the ancient mediterranean, 2nd edn. Oxford University Press, Oxford, UK

Garbrecht G (1980) The water supply system at Tuspa (Urartu). World Archaeol II(3):306–312

Graham JW (1987) The palaces of Crete. Princeton University Press, Princeton, NJ

Jansen M (1989) Water supply and sewage disposal at Mohenjo-Daro. World Archaeology, The Archaeology of Public Health 21(21):177–192

Javan M, Hassanli AM, Shahrokhnia MA (2006) The ancient qanats of Iran. In: Angelakis AD, Koutsoyiannis D (eds) Proceedings 1st IWA International Symposium Water and Wastewater Technologies in Ancient Civilizations, National Agricultural Research Foundation, Iraklio, Greece, held 28–30 October 2006, pp. 531–534

Knauss J (1991a), Mykenische Talsperren in Arkadien and Beotien. Historische Talsperren 2: 19–41

Knauss J (1991b), Arkadian and Boiotan, centres of Mycenaean hydraulic engineering, Irrigation and Drainage, vol. 5. Kluwer Academic Publishers, Netherlands, pp. 363–381

Koutsoyiannis D, Zarkadoulas N, Angelakis AN, Tchobanoglous G (2007) Urban water management in Ancient Greece: Legacies and lessons. J Water Resour Planning Manage, ASCE, 134(1)

Laureano P (2001) The Water Atlas: Traditional Knowledge to Combat Desertification Copyright by Peatro Laureano, Translated by Anna Cirella and Angela Whitehouse

Mays LW (2006) Water sustainability: parallels of past civilizations and the present. In: Mays LW (ed) Water resources sustainability. McGraw-Hill, New York, USA

Mays LW, Koutsoyiannis D, Angelakis AN (2007) A brief history of urban water supply in antiquity. Water science & technology: Water Supply, IWA Publishing 7(1):1–12

Mays LW (2008) A very brief history of hydraulic technology during antiquity. Environ Fluid Mech 8(5):471–484

Moosavi SA (2006) Qanat invention puzzle. In: Angelakis AD, Koutsoyiannis D (eds) Proceedings 1st IWA International Symposium Water and Wastewater Technologies in Ancient Civilizations, National Agricultural Research Foundation, Iraklio, Greece, held 28–30 October 2006, pp. 535–540

Myers WJ, Myers EE, Cadogan G (1992) The aerial atlas of ancient Crete. University of California Press, Berkeley and Los Angeles, CA

Oleson JP (1984) Greek and Roman mechanical water-lifting devices: the history of a technology. University of Torronto Press, Torronto

Oleson JP (1995) The origins and design of Nabataean water-supply systems. Studies in the History and Archaeology of Jordan, vol V, pp. 707–719, Department of Antiquities, Amman, Jordan

Ortloff CR (2005) The water supply and distribution system of the Nabataen city of Petra (Jordan, 300 B.C. A.D. 300. Cambridge Archaeological J 15(1):93–109

Ortloff CR, Crouch DP (1998) Hydraulic analysis of a self-cleaning drainage outlet at the Hellenistic City of Priene. J Archaeological Sci 25:1211–1220

Parsons AW (1943) Klepsydra and the paved court of the Pythion. Hesperia 12:191–267

Papademos DL (1975) The hydraulic works in ancient Greece, Vol. B. (in Greek), Ed. TEE, Athens.

Ruben I (ed) (2003) The Petra Siq: Nabataen hydrology uncovered. Petra National Trust, Amman, Jordan

Sklivaniotis M, Angelakis AN (2006) Water for human consumption through the history. In: Angelakis AD, Koutsoyiannis D (eds) Proceedings 1st IWA International Symposium Water and Wastewater Technologies in Ancient Civilizations, National Agricultural Research Foundation, Iraklio, Greece, held 28–30 October 2006, pp 659–666
Tasios TP (2002) From "Peisistratean" to Evinos (in Greek), Kathimerini
Taylour WW (1983) The Mycenaeans. Thomas & Hudson
Ur J (2005) Sennacherib's northern Assyrian canals: new insights from satellite imagery and aerial photography, Iraq LXVII, pp. 317–345
Van der Waerden BL (1968) Eupalinos and his tunnel. ISIS 59(1):82–83
Viollet PL (2003) The predecessors of European hydraulic engineers: Minoans of Crete and Mycenaeans of Greece (2100–1200 B.C.). Proceedings XXX IAHR Congress, Theme E: Linkage Between Education Research and Professional Development in Water Engineering, Thessaloniki, Greece, August
Viollet PL (2006) Water management in the early bronze age civilization. Proceedings, La Ingenieria Y La Gestion Del Agua a Traves de Los Tiempos, held at the Universidad de Alicante, Spain, with the Universidad Politechnica de Valencia, Spain
Viollet PL (2007) Water Engineering in Ancient Civilizations 5000 years of History, International Association of Hydraulic Research (IAHR), originally in French, L'Hydraulique dans les Civilisations Anciennes, translated into English by Forrest M. Holly, Jr
Zangger E (1994), Landscape changes around Tiryns. Am J Archaeol 98:189–212

Bibliography

Adeel Z, Schuster B, Bigas H (2009) What makes traditional technologies tick? A review of traditional approaches for water management in drylands, UNU Desertification Series No. 8, United Nations University, International Network on Water, Environment and health
Cadogan G, Clarke P (1991) Palaces of Minoan Crete. Routledge
Castleden R (1990) Life in Bronze Age Crete. Routledge
Coopey R, Fahlbusch H, Hatcho N, Jansky L (2005) A history of water issues: lessons to learn. United Nations University, Toyko, Japan
Evans AJ (1921) The Palace of Minos at Knossos, vol. 1. Macmillan, London, UK
Evans AJ (1935) The Palace of Minos at Knossos, vol. 4. Macmillan, London, UK
Gibson D (2003) The Nabataeans, builders of Petra, CanBooks, Osler, SK, Canada
Gorokhovich Y (2004) Abandonment of Minoan palaces on Crete in relation to the earthquake induced changes in groundwater supply. J Archaeological Sci 32(2):217–222
Graham JW (1987) The palaces of Crete. Princeton University Press, Princeton, NJ
Halstead P (1997) Storage strategies and states on prehistoric Crete: reply to Strasser (JMA 10, 1997, pp. 76–100) J Mediterranean Archaeol 10:103–107
Hutchinson RW (1950) Prehistoric town planning in Crete. Town Planning Rev 21:199–220
Mays LW (2002) Urban water infrastructure: a historical perspective. In: Mays LW (ed) Urban water supply handbook. McGraw-Hill, New York, USA
Platon N (1971) Zakros: The discovery of a lost palace of ancient Crete. Charles Scribner's and Sons, New York
Strasser TF (1997) Storage and states on prehistoric Crete: the function of the koularouras in the first Minoan palace. J Mediterranean Archaeol 10.1:73–100
Watrous LV, Hatzi-Vallianou D, Blitzer H (2004) The plain of Phaistos: Cycles of complexity in the Messara region of Crete, Momenta Archaeologica 23. Cotsen Institute of Archaeology, University of California, Los Angeles

Chapter 2
Water Technology in Ancient Mesopotamia

Aldo Tamburrino

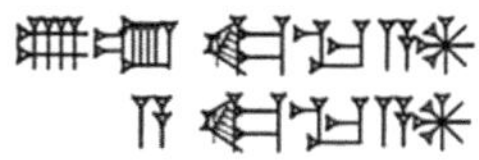

ú-um inim-ma-àm
a inim-ma-àm
Food is the matter
water is the matter

Sumerian Proverb

2.1 Introduction

Mesopotamia is in the east side of the region named "fertile crescent", where agriculture flourished and the earliest civilizations were born more than eight thousand years ago. In the alluvial plain of Lower Mesopotamia agriculture based on irrigation developed, in contrast to the Upper Mesopotamia, where dry-farming was possible. A complex system of canals and waterworks developed, with the dual function to ensure irrigation and to be used as waterways. Control of water was decisive as a way to guarantee economic prosperity, but also was a source of inter-state conflicts and a political tool. Water technology was not limited to irrigation, Mesopotamians also pioneered in sanitary engineering, with many cities presenting networks of wastewater and stormwater drainage systems. Overexploitation of land and water resources for agriculture affected the environment, resulting in silting and soil salinization, matter that has been recorded since the earliest cuneiform writings.

Mesopotamia is the name given by the ancient Greeks to the land lying between the Euphrates and Tigris rivers and its tributaries, roughly comprising modern Iraq and part of Syria (Fig. 2.1). Considering that the origin and development of ideas and techniques are difficult to restrict to a well defined geographical area because migration, trade and military operations contribute to their diffusion, the term "Greater

A. Tamburrino (✉)
Department of Civil Engineering, University of Chile, Casilla, 228-3, Santiago, Chile
e-mail: atamburr@ing.uchile.cl

L.W. Mays (ed.), *Ancient Water Technologies*, DOI 10.1007/978-90-481-8632-7_2,

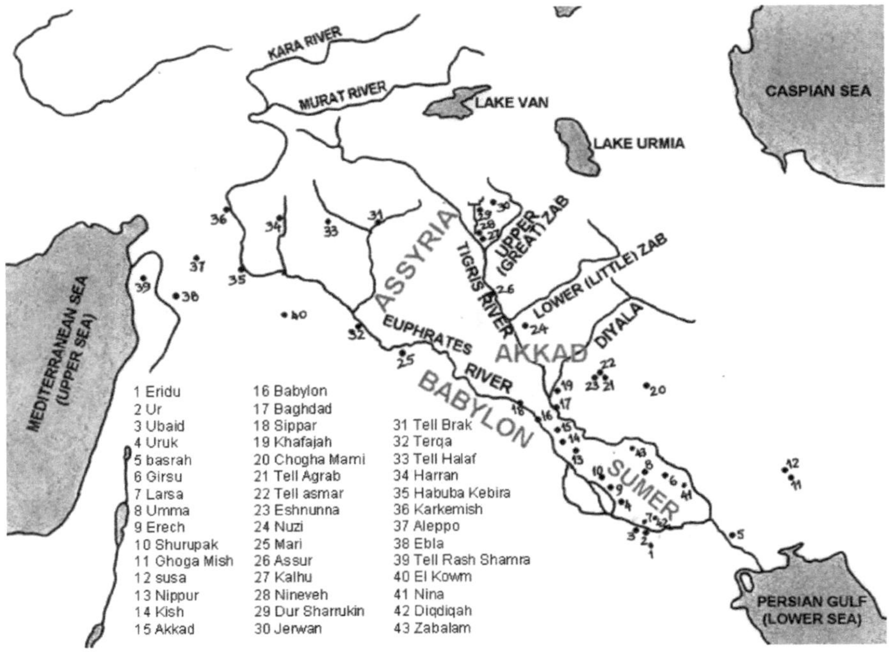

Fig. 2.1 Greater Mesopotamia. (Drawn by the author from several sources. Locations are approximate)

Mesopotamia" has been devised, including western Iran, eastern Syria, and south-eastern Turkey as part of the sphere of influence of "the land between rivers". Although in the text we will refer to Mesopotamia, it has to be understood in this broader sense.

The Euphrates and Tigris rivers have their source in Armenia. The Euphrates (2,800 km long) results from the junction of two branches, the Kara (western Euphrates), and the Murat (eastern Euphrates), northwest of Lake Van, following a zigzagging course in a general south-west direction in Turkey, and changing to a south-east direction across the Syrian plateau and Iraq. The Tigris (1,950 km) has its source near the Mount Ararat and almost immediately flows south-east. When the Euphrates and Tigris emerge from the Taurus Mountains, the two rivers are separated from each other about 400 km. Near Baghdad they are just 32 km apart, but they soon diverge again to join at Qurnah, 100 km north of Basrah, to form the Shatt-el-'Arab (River of Arabia), which flows for about 220 km before reaching the Persian Gulf. Mesopotamia is part of what Breasted (1916, p. 101) called 'the fertile crescent': '. . .approximately a semicircle, with the open side toward the south, having the west end at the southeast corner of the Mediterranean, the center directly north of Arabia, and the east end at the north end of the Persian Gulf. . .', where agriculture flourished and the earliest settlements took place. It is usually divided into Upper and Lower Mesopotamia. Upper Mesopotamia is a large piedmont zone flanked by semiarid highlands where dry-farming is possible. Lower Mesopotamia is an alluvial plain where irrigation-based agriculture developed. In this region, the Tigris and Euphrates flow with such a low gradient that they meander considerably and throw numerous side-branches. Like all meandering rivers they raise their own beds, so that they frequently flow above the level of the plain, their overflow tending to create permanent lakes and marshes, and they occasionally change their course (Roux, 1992). As far as 500 km from the Gulf, the average slope of the terrain is around 1:25,000. Six and five thousands years ago, the shoreline was situated more than 200 km farther inland than it is nowadays, and cities like Ur, Eridu and Diqdiqah were virtual seaports. The marshes south of these cities were considered by the ancients as part of the Persian Gulf (Jacobsen, 1960), which was called the 'Sea' (or 'Lower Sea' to distinguish it from the Mediterranean, that was the 'Upper Sea', King, 1918, p. 8).

Mesopotamian cosmography is deeply linked to the water, and in many aspects is a mirror of the geography known by the people that inhabited the region. Thus, in the "Babylonian map of the world" (a map drawn in a clay tablet from the Late Babylonian period, no older than the ninth century B.C., Horowitz, 1998), the earth is surrounded by the cosmic ocean, beyond which there exists several (eight?) regions, depicted as triangles in the map, and have been interpreted as islands, mountains or distant land masses (Horowitz, 1998, pp. 30–32). In the Enuma Elish, the Babylonian Epic of the Creation, Marduk (a former second rank god that became the most important of the Babylonian pantheon under Hammurabi's reign) created the sources of fresh water in both heaven and earth. First, he produced precipitation and appointed himself as the god controlling the weather (Horowitz, 1998, pp. 117–118): '... He collected it and rolled into clouds / To raise the wind, to make the rain fall ...'. After creation of precipitation, Marduk created the sources of fresh water on the earth from the body of the goddess Tiamat: '... He set up her head, headed up dirt / Then he opened up the string, it became saturated with water / Then he opened the Euphrates and Tigris in her eyes / He plugged her nostrils, left ... behind /He heaped up the distant mountains on her breast / Then he drilled a water hole to carry the catchwater...'. According to Horowitz (1998), 'distant mountains seem to include all the mountains under the Sun, and the waters of the water hole and catchwater presumably refer to pools of water in the mountains that feed the springs that flow down to the plains'. It is interesting to mention the role of the mountains as source of fresh water in this early understanding of the world's hydrology.

The Euphrates (Sumerian: Buranun; Akkadian: Purattu) and Tigris (Sum.: Idigina or Idigna; Akk.: Idiqlat) rivers shaped both the society and the mind of ancient Mesopotamians. The climate of central and southern Mesopotamia is such that flood periods of the rivers occur between April and June, too late for winter crops and too early for summer crops. For an agriculture based society, to ensure irrigation was a permanent concern, and a complex system of canals, reservoirs, dykes, etc. was developed in order to ensure appropriate water supply to the fields (Roux, 1992). Although the concept of 'hydraulic society' (in the sense of oriental despotic power, Wittfogel, 1957) has been challenged and evidence exists of irrigation systems used in the early stage of the state formation, previous to the existence of a powerful centralized state (see, for example, Adams, 1974; Gibson, 1974). The fact is that large labor forces and cooperation of many communities were necessary to create and maintain an efficient network of canals. Independently of the discussion regarding the sequence of the formation of the state and the building of irrigation nets, there is no doubt that, at some point, construction and maintenance needed some centralized control. Thus, it is not a surprise that one of the laws from the first written law-"code", the laws of Ur-Nammu (first king of Sumer and Akkad between 2112 and 2085 B.C.), refers to the penalty that a man has to pay when he causes to flood a cultivated field. Law 28 says: 'If a man caused water to flood the cultivated field of a(nother) man, he shall measure out (to him) three *gur* of barley per *iku* of field'[1] (Finkelstein, 1966). The same idea is reinforced 350 years later in the Code of Hammurabi (1795–1750 B.C., king of Babylon), where several laws deal with maintenance and operation of dykes and canals (Harper, 1904).[2] Mesopotamia was also the place where writing was invented in a date earlier than 3000 B.C., evolving from clay tokens used to record concrete quantities up to the cuneiform writing, able to document elaborated thoughts (Schmandt-Besserat, 1992).

Water was ever present in all aspects of Mesopotamian life, including religion, politics, law, economy, international affairs, war, etc. It cannot be overlooked that the word for water in Sumerian is the same for semen ("a", written 𒀀). The Enuma Elish indicates that in the beginning already existed the waters,[3] distinguishing between Apsu, conceived as a male god of sweet waters, and Tiamat, his spouse, a goddess of salt water. Also, apsu denotes the freshwater upon which the earth floated. As underground waters, apsu may be reached when laying the foundations of a temple,[4] and also appears naturally in pools and marshes (Jacobsen, 1946). The apsu is the domain of one of the most important gods in the Mesopotamian pantheon, Enki

---

[1] 1 *gur* = 0.3 m^3, 1 *iku* = 3600 m^2 (0.36 ha).

[2] Law 53: If a man neglect to strengthen his dyke and do not strengthen it, and a break be made in his dyke and the water carry away the farm-land, the man in whose dyke the break has been made shall restore the grain which he has damaged.

Law 55: If a man open his canal for irrigation and neglect it and the water carry away an adjacent field, he shall measure out grain on the basis of the adjacent fields.

Law 56: If a man open up the water and the water carry away the improvements of an adjacent field, he shall measure out ten *gur* of grain per *gan*. (1 *gan* = 27,000 m^2 = 2.7 ha).

[3] When in the height heaven was not named,
And the earth beneath did not yet bear a name,
An the primeval Apsu, who begat them,
And chaos, Tiamat, the mother of them both, –
Their waters were mingled together, (King, 1902).

[4] From a soil mechanics point of view, to reach the water table when lying the foundations of a building is the right solution for saline soils. Dry saline soils have a load capacity much higher than if they are wet. Thus, a soil that goes through dry – wet cycles has a variable load capacity, inducing differential settlements and damaging the structure. If foundations reach the water table, they lie in a soil where salt dilution is maximum, with a more demanding construction condition. This analysis is valid when baked bricks are used in the foundation.

Fig. 2.2 Seals with the representation of the water god Enki – Ea. *Upper image*: Old Babylonian seal (ca. 1700 B.C.). The god with water flowing from his shoulders. *Lower image*: Akkadian cylinder seal (ca. 2300 B.C.). The god seated on a stool and holding up a jar from which water spouts and falls. (Formerly in the private collection of Dr. Bron Lipkin, London, UK)

(Sum.) -or Ea (Akk.)- who is depicted with a cascade of water emanating from his shoulders, or holding a vase from where water emerges, as shown in Fig. 2.2. Enki, as master of the fresh water is a creator god. It should not be overlooked that Enki is a wise god, always ready to help humans.

Mankind was created to alleviate the gods from the hard work they had to do. Tablet I from the Atra-Hasis epic about the creation and early history of man begins[5]: 'When the gods like men / bore the work and suffered the toil...'. The heavy work included digging and maintenance of canals. But the lesser gods did not tolerate this state of affairs, and they rebelled ('... they set fire to their tools, / fire to their spades they put / and flame to their hods ...'). As a result, mankind was created: '... Let the birth-goddess create offspring / and let man bear the toil of the gods ...'. And men 'with picks and spades they built the shrines, / they built the big canals banks / for food of the peoples, for sustenance of the gods'. However, 'Twelve hundred years had not yet passed', the peoples multiplied and with their

[5]The cites from Atra-Hasis epic are from Lambert and Millard (1999)

noise disturbed the gods who decided to exterminate mankind. At this point, Enki intervened alerting one man, Atra-Hasis, who was instructed how to escape from the plan of the gods and let the mankind to survive. Tablet II refers to one of the ways the gods attempted to eradicate mankind from the earth: '... Adad [god of weather] should withhold his rain, / and below, the flood should not come up from the abyss / .... / let the fields diminish their yields ...'. Famine arose not only from the drought, but also because the soil turned no apt for agriculture due to salinization: '... the black fields became white, / the broad plain was choked with salt ...'. Not succeeding with famine (due to Enki's intervention), the gods decided a devastating flood that 'tear up the mooring poles' and 'make the dykes overflow'. Tablet III describes that 'for seven days and seven nights / came the deluge, the storm, the flood'. The Sumerian version of the flood (Lambert and Millard, 1999) explicitly says that construction and maintenance of water channels was a god's decision '... he (the god) did not stop the yearly flood, but dug the ground and brought the water, / he established the cleaning of the small canals and the irrigation ditches ...'.

The goal of artificial canals was not only to ensure irrigation, but also they were used as waterways. An example is found in Gudea's cylinder A, mentioning that Gudea, king of Lagash between 2141 and 2122 B.C., set the foot on his cargo boat, headed it towards the city of Nina by the Idninashdu, navigating with joy along the "new channel" (Lara Peinado, 1996). The Idninashdu was an important artificial canal, fed with water from the Euphrates, connecting the cities of Nina (current Surghul), Lagash, Girsu and Zabalam. Representations of boats are found in seals, relieves and models, like those depicted in Figs. 2.3, 2.4, and 2.5.

The double function of artificial canals was not exempt of stress, particularly for those officials responsible of agricultural production before the ruler. As Bonneterre (2003) pointed out, to ensure navigability in artificial canals, it was necessary to raise the levees, increasing the hazard of a failure, and risking not just the canal but also the cultivated area.

Fig. 2.3 Impression of an Akkadian (2350–2000 B.C.) cylinder seal with a boat scene. (Louvre museum. Photo taken by the author)

Fig. 2.4 Detail of a bass relief from the palace of Sargon II (Louvre museum. Photo taken by the author)

Fig. 2.5 Model of a boat found in Jemdat Nasr (ca. 3000 B.C.) (Ashmolean Museum, University of Oxford)

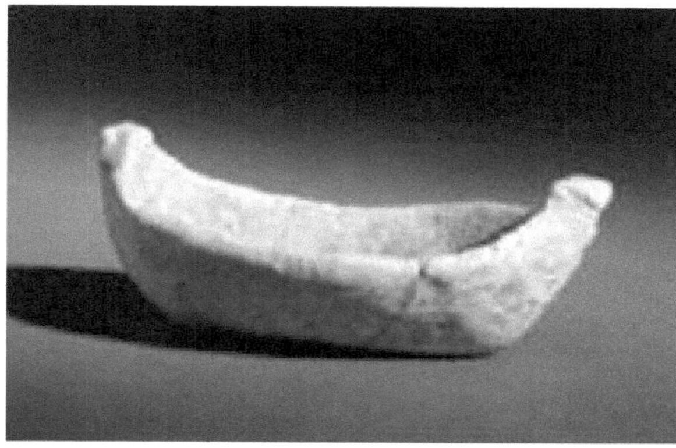

2.2 "Applied Hydraulics"

The reason of developing a hydraulic knowledge by Mesopotamians was a practical one and it is well condensed in the Sumerian proverb cited at the beginning of the chapter. It was the natural consequence of a necessity in a given environmental context. It is not required to be aware of the physical principles governing hydrostatics and hydrodynamics to take advantage of them. As examples of "applied elementary physics" we can mention the flow induced by a pressure difference and buoyancy, both cases represented in Mesopotamian art. Figure 2.6 presents the impression of a cylinder seal with two persons drinking beer from a vessel using a straw. Copper tubes used as "straws" have been found. They have small perforations at the lower end to avoid barley chaff from entering the tube. Figure 2.7 shows inflated skins used as buoyant elements. Both images are details of stone panels from the palace of Ashurnasirpal II (ca. 910 – 859 B.C., king of Assyria from 883 to 859 B.C.). The first image represents two men swimming in a river helped by inflated skins. The second describes transportation of large stone blocks in a raft composed of inflated skins. Sealing and impermeability was achieved using bitumen. Another device,

Fig. 2.6 Impression of a cylinder seal showing two persons drinking beer using copper tubes as "straws". Sumerian, Early Dynastic (ca. 2700 B.C.). (Martin, 1940, Plate 2)

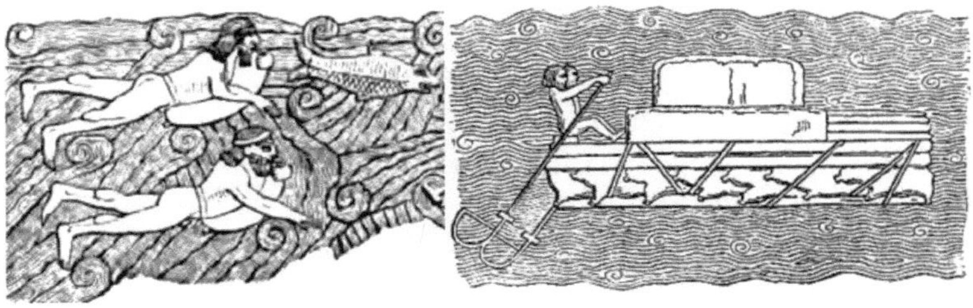

Fig. 2.7 Inflated skins used as bouyant elements. Details of stone panels from the palace of Ashurnasirpal II (ca. 910–859 B.C.) Images taken from Layard (1851, p. 220) and von Reber (1882, p. 93)

mentioned several times in a couple of tablets in the British Museum,[6] is the water clock or clepsydra (^{giš}dib-dib 𒄑𒁳𒁳, Sum. *dibdib,* Akk. *dibdibbu*). Those tablets are dated from the Kassite period (second half of the second millennium B.C.) or older. Water-clocks played an important role as an aid to astronomical calculations. Time was measured by the volume or weight of the water discharged rather than level difference (Nemet-Nejat, 1993, pp. 70–72). When Enki reveals to Atra-Hasis the imminent flood the legend says that the god announced the time: "He opened the water-clock and filled it; He announced to him the coming of the flood for the seventh night" (Lambert and Millard, 1999; Tablet III, lines 36–37).

Although still is an open discussion, some bass reliefs in the palace of Sennacherib (king of Assyria, 705–681 B.C.) at Nineveh and literary references

[6]Tablets first published by King (1900) *Cuneiform Texts from Babylonian Tablets in the British Museum 9*, and analysed by Nemet-Nejat (1993), pp. 70–72.

Fig. 2.8 Shadufs. *Upper image*: Bass relief in Sennacherib's palace, after Layard (1882, p. 25) *Lower image*: Impression of an Akkadian seal (end of the third millennium B.C.). Drawn by the author after an image from Boehmer (1965, Table LX, seal 716)

would suggest that the water screw was used in Mesopotamia several centuries before Archimedes (287–212 B.C.) (Dalley, 1993; Dalley and Oleson, 2003). Another bass relief in Sennacherib's palace shows water lifting by means of *shadufs* (Fig. 2.8), ([cuneiform], zi-ri$_2$-gum$_2$, or [cuneiform], zi-ri$_2$-gum. Sum. *zirigum*, Akk. *zirīqu*) which were used in Mesopotamia as early as the third millennium B.C. According to the royal chronicle of Lagash, the bucket needed to build the *shaduf* was a gift of the gods An and Enlil to the people.[7]

When Ashurnasirpal II moved the capital of the Assyrian Empire from Assur to Kalhu he had to build a canal taking waters from the Upper Zab River to irrigate the fields and the palace gardens. Later, Sargon II, "king from Assyria and Babylon" between 722 and 705 B.C., founded a new capital, Dûr-Sarrukin ("Fortress of Sargon", near Khorsabad), embellished with gardens. However, their engineering and hydraulics achievements were surpassed by Sargon's son, Sennacherib, king between 705 and 681 B.C., who decided to move the capital to Niniveh, a decision that was accompanied by an intense engineering activity, directed by himself, as recorded in an octagonal cylinder of baked clay (British Museum, 1909), in which he claims to 'have knowledge of all handicraft'. He built a big palace 'that it might be a wonder to behold among hosts of mankind', which he called 'The Palace that has no rival', where 'every day water for irrigation might flow in abundance' by means of 'levers of bronze, and buckets of bronze, and in place of the draw-wells

[7] '. . . To dig the canals, to dredge the irrigation dredges, to irrigate with the shadouf the vast campaigns, to use abundant water to wet the meadows and fields, (An and Enlil) put at disposal of the people the spade, the hoe, the bucket, the plough that give life to the land . . .' (Translated from Glassner, 1993, p. 152)

great beams and wooden frame-works over the well-shafts' (Column VII, lines 45–49). No doubt he refers here to *shadufs*. Niniveh was enlarged, two surrounding walls were built, gardens above and below the city were laid out, and a complete water supply system was designed and erected ('... in order to increase the planting, from the district of Kisiri, through high ground and low ground with pickaxes I dug, I directed a canal; those waters I set over against the neighborhood of Nineveh, and I led them among the orchards by means of irrigation channels...', Column VIII, lines 25–30). Water from springs and pools in Mount Musri, properly collected, increased the discharge of the Kosh River. However, it was not enough, and in 690 B.C., a 50 km long channel-aqueduct, with a portion as a tunnel, was finished (it was built in 13 months, an achievement even in our days) (Jacobsen and Lloyd, 1935). This system had its 'mouth' at Bavian, north-east from Nineveh, and crossed a valley by means of an aqueduct 280 m long built up on stone arches to pass through a 'deep ravine', as Sennacherib called the depression crossed by the aqueduct at Jerwan (Jacobsen and Lloyd, 1935, p. 6). The aqueduct rested on six piers (about 7 m high from the bed level), properly protected with breakwaters. The canal was 22 m wide and 2 m high. As Dalley and Oleson (2003, p. 6) say, Sennacherib was 'particularly pleased to have designed an automatic sluice gate that opens by itself, without using a spade or shovel ... (it) is not opened by any action of men's hands'. Sennacherib was able to ensure, throughout the year, enough water to irrigate the gardens, orchards, parks and plantations he had built at Nineveh. However, spring was a period potentially dangerous because the floods generated by the snow melting on the mountains. He solved the problem using an outlet that would divert the flood to a marsh specially designed for that purpose: 'To bring the course of those water to rest I created a swamp, and a reed plantation within it I planted. *Igirû*[8]-birds, wild swine, and ... of the forest I let loose therein' (Column VIII, lines 46–49, British Museum, 1909). This seems to be the first written description of retaining ponds to regulate floods, designed with an environmentally friendly perspective.

2.3 Sanitary Engineering

Water technology development in Mesopotamia and its neighborhood is usually associated with irrigation, but not less surprising is the level of sophistication reached in sanitary engineering, including wastewater facilities and stormwater drainage systems. There is evidence that, as early as 6500 B.C., there was a well developed urban settlement at El Kowm 2, Central Syria, about 80 km south from the Euphrates. The city had well planned houses, many of them with drainage systems for domestic wastewater (Stordeur, 2000). Later, around 3500 to 3000 B.C.,

[8]Igirû: Heron. A swamp bird, but exclude the stork. (Oriental Institute 2004).

many cities of Mesopotamia had networks of wastewater and stormwater drainage. Habuba Kebira, now under the waters of Lake Assad created by the Tabaqah Dam, was a Sumerian colony in the margins of the Euphrates, north of El Kowm. It had an important although short existence, located in the middle of the route to the forest and mineral resources in Turkey and northern Syria. It was created with the specific purpose to serve as a trading post and did not evolve from previous settlements. Thus, the city '... was given a true urban layout, the first one in history for which we have evidence ... [with a hierarchized] road system provided with a net of drain pipes...' (Vallet, 1997). The plan of the city comprised about 12 ha, with all the streets paved with alluvial gravel. 'But most spectacular in the roadway network is its system of canalizations' (Vallet, 1997, p. 72). According to several sources cited by Vallet, houses usually had several drains to evacuate wastewater and rain. Sewage disposal could be achieved in three ways: directly to the gutter or canalization (canal) of a street; towards pits located in the city if the house was close to one of them; and for the houses being next to the city walls, directly out of the city by means of canals regularly spaced along the ramparts. The streets had a vast system of interconnected canals which, following the natural slope of the terrain carried the wastewater and rain to the countryside, outside of the city walls. There existed three types of canals: those having the side walls and cover made from limestone slabs, with a bottom formed by two or three layers of compacted clay, a second type consisting in U-shaped open drains, made of clay in units 64 cm long, and a third type made of joined clay pipes (Ludwig, 1977, cited by Vallet, 1997).

It is evident that the people that traced the layout of Habuba Kebira were not designing for the first time a city drainage system, but incorporating the knowledge, techniques and designs of already existing systems in the metropolis of Uruk, situated 900 km from the colony. It is natural and expected, as urbanism spread, that these complex drainage systems should have become quite common in Mesopotamian cities from the end of the fourth millennium and beginning of the third, B.C. Vallet (1997) concludes that, the creation of urban drainage systems is 'an asset of the Uruk culture and the question is to determine the reasons for which it was apparently lost thereafter'.

East of the Tigris River, in current southwestern Iran, between the Dez and Karum Rivers, about 75 km the place where they join, Chogha Mish, flourished. The region is apt for dry farming, and major channel irrigation does not seem to have been practiced until around 1500 B.C. (Alizadeh, 2008). However, as early as the Protoliterate period (around 3400–2900 B.C.) it was a "planned town with streets, side alleys, sewer and irrigation drains, water wells and cesspools, workshops, and public and private buildings" (Alizadeh, 2008, p. 26). Drainage systems were formed by clay pipes and baked clay bricks, as shown in Figs. 2.9 and 2.10

The Diyala River is a major tributary of the Tigris River in its basin. The Oriental Institute of the University of Chicago participated in several campaigns between 1930 and 1938 aimed to excavate private houses in the Diyala region (northeast

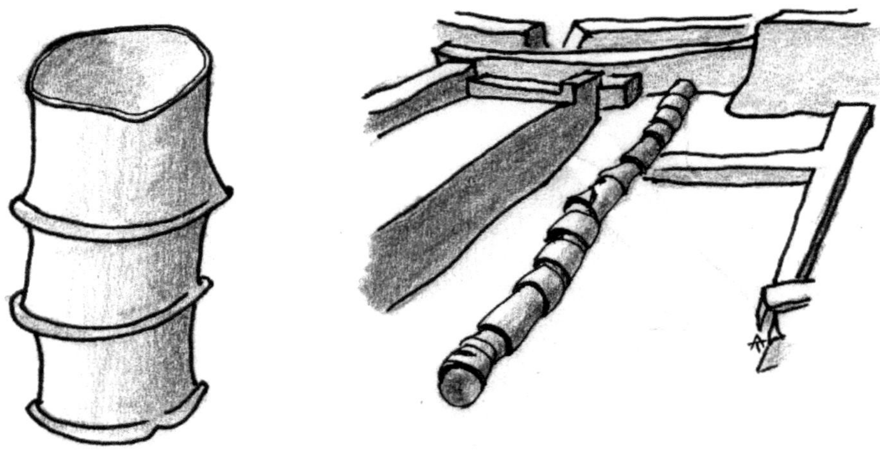

Fig. 2.9 Drain pipes found in Chogha Mish. *Left image*: Old Elamite period (about 2700–1600 B.C.) *Right image*: Protoliterary period (about 3400–2900 B.C.). Drawn by the author after photographs in Alizadeh (2008)

Fig. 2.10 Drainage canalization made from baked clay bricks. Chogha Mish. Protoliterary period (about 3400–2900 B.C.). Drawn by the author after photographs in Alizadeh (1996)

of Baghdad), and the results of the expeditions at Khafagah, Tell Asmar and Tell Agrab were presented in a report in 1967 (Delougaz et al., 1967). These houses range from about the middle of the Protoliterate period (3300 B.C.) to around 1800 B.C. Clay tablets with drawings of plans of houses were found. In the concluding remarks of the report is established that '. . . In the temples baked bricks and bitumen were freely used for pavements, drains, and other constructions involving the use of water, but the use of these relatively expensive materials was rather limited in private houses. Toilet facilities were found in a few houses of the Akkadian (2335–2155 B.C.) and later periods at Tell Asmar, but numerous structures which resembled toilets were built in both the Early Dynastic III (2600–2350 B.C.) and the Protoimperial (3000–2850 B.C.) [periods]. . .'. Figure 2.11 shows a toilet whose seat was coated on top with bitumen and 'built up of baked bricks (37.5×37.5×7.5 cm), five courses high, with a slot of 10 cm wide through the middle. Below the slot, in the paved floor, was a hole about 12 cm square beneath which was carefully fitted a baked-clay pipe with an opening 12 cm in diameter. The joint between the mouth of the pipe and the bricks around the square hole was smoothly calked with bitumen. The pipe was set into an opening in the covering of the uppermost section of the vertical baked-clay drain found below the toilet. The drain was 57 cm in diameter, each of its four sections being 32 cm long.' (Delougaz et al. p. 176). In the figure, an elevation showing the fitting of the toilet to the vertical drain is also given. Example of a vaulted sewer canal made of baked bricks is presented in Fig. 2.12.

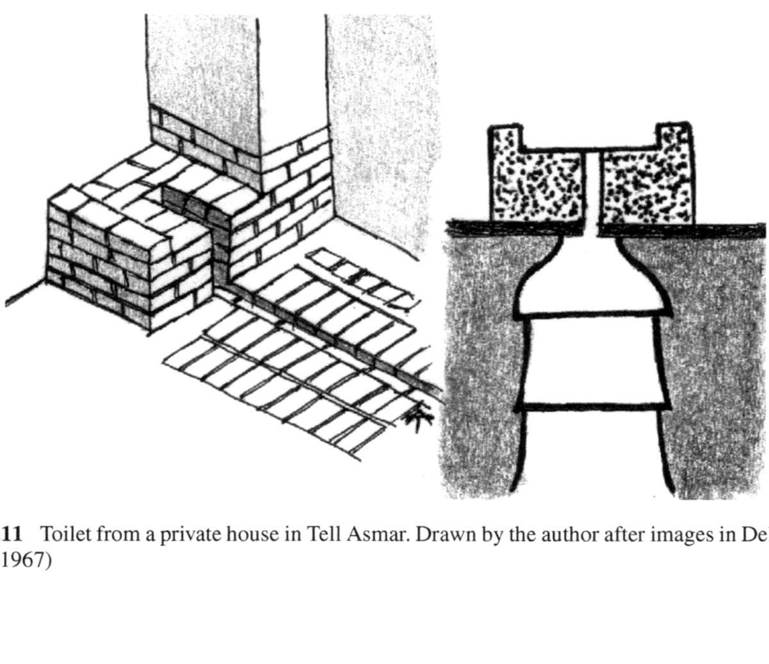

Fig. 2.11 Toilet from a private house in Tell Asmar. Drawn by the author after images in Delougaz et al. (1967)

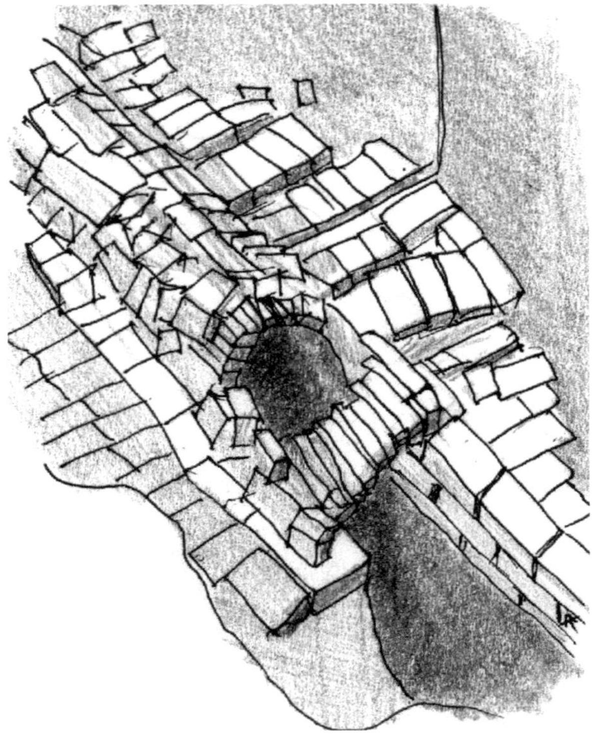

Fig. 2.12 Vaulted sewer of baked bricks below a street of Tell Asmar. Drawn by the author after a photograph in Delougaz et al. (1967)

2.4 Irrigation

There are no records of canal design, at least in the sense we understand it now. Tablets record mainly the manpower and time to perform canal or irrigation works, and computations related to digging and brick wall works, but they are more mathematical exercises than minutes of actual man work. In this regard, there were two classes of people: the scribes, and the laborers. The former included surveyors who would use measuring rods (reeds in fact) and lines, and would have a lot of calculation to do in connection with leasing out plots of irrigated land, and of seeing that the system of canals and irrigation ditches was kept in order. But the actual work of earth moving was certainly done by illiterate laborers under foremen. The surveyors no doubt gave details of, e.g., a canal so wide and so deep, and they then calculated how much earth was moved in so many days by so many men. But this calculation was all very academic: obviously no one actually measured the quantity of earth which the laborers dug out of so many cubits length of canal so wide so deep: the realities of the job would not allow that.

Control of water was decisive as a way to ensure economic prosperity and, as such, as a political tool and a source of inter-state tensions, as reflected, for example in the Umma-Lagash conflict. But the proverb 'Can strong warriors resist a flood?' (Lambert, 1996, p. 266), written in a bilingual tablet from the Late Babylonian period (first millennium B.C.) and transmitting a much older tradition, indicates that water was also a tool of war, used to stop or obstruct the movement of enemy armies. As an example, we can mention the action by which Abi-eshuh, king of Babylon between 1711 and 1684 B.C., planned to capture Iluma-ilu.[9] Although Abi-eshuh's campaign was not successful, one of the year names (the nineteenth) of his reign is 'The year that Abi-eshuh the king, by the exalted might of Marduk, dammed the Tigris',[10] indicating how he wanted to defeat the enemy army. Notably, there is a tablet describing the way that Abi-eshuh diverted the Tigris River, translated by Lambert (2007, p. 57). It is not an administrative text, as we would expect, but and oracle document. Since the answer could only be "yes" or "no", all details of the plan had to be in the questions. Thus, a hydraulic work used to dam the river and divert the water to a large lateral irrigation canal can be identified. It is named *bīt turri*, term that Lambert translated as 'house of turning' (Lambert 2007, p. 150). Basically, it consists in two gates that can turn, blocking the river or the entrance to the canal depending on their positions. Probably they were made of reeds and bitumen or, also, wood (Durand, 1990, p. 135). According to the oracle text, they did not block the river completely because the question 'by [heaping up] reed bundles and earth should they dam [the river]?' is asked. Thus, turning the gates toward the Tigris and adding some material to close the breach between them, Abi-eshuh forced the water into the irrigation system and inundated the fields, making difficult the movement of enemy forces. Figure 2.13 shows the author's conception of the structure, after the descriptions by Lambert (2007) and Durand (1990).

Sumerian language is full of words related to water, canals, hydraulic structures, etc. Although the exact meaning of many words is not known, several studies have been published looking for precise explanation of the terms related to canals and

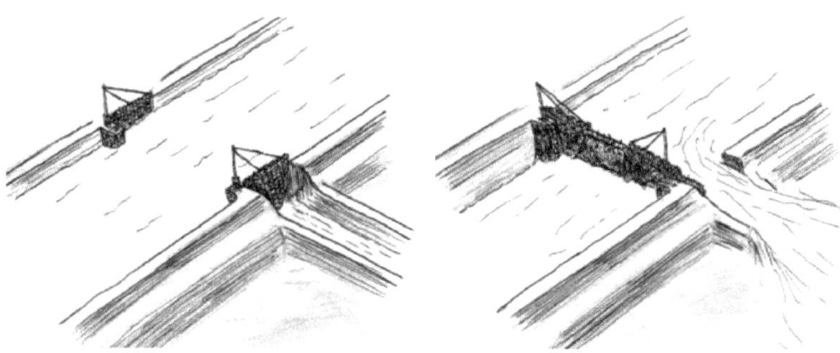

Fig. 2.13 Damming of the Tigris by Abi-eshuh according to the oracle questions

[9]King of the "Country of the Sea" who was challenging the power of Babylon. The "Country of the Sea", as it is called in the Babylonian Chronicles was in the shores of the Persian Gulf and probably included Uruk.

[10]Years were named after some important event or action happened during the reign of the king.

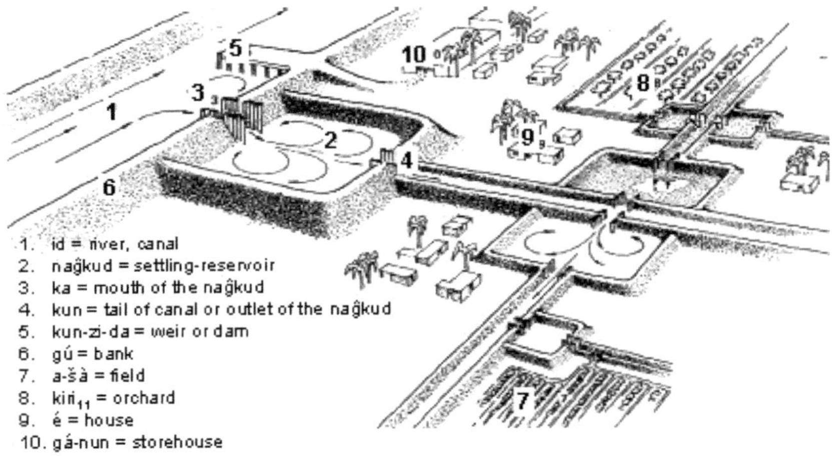

Fig. 2.14 The settling-reservoir (*nag-ku*$_5$) and complementary water-works, as reconstructed by Kang (1973) from the Ur III texts

water-works. Thus, Kang (1973) studying the term nag-ku$_5$ (*naĝkud*, 𒉀𒆪), was able to propose an image of Sumerian canal and irrigation systems, arriving to a description of a multi-purpose settling-reservoir, as shown in Fig. 2.14, serving 'to facilitate intersections, to slow water flow from higher to a lower plane, to prevent scouring and erosion, and to act as a kind of a simple reservoir'. However, according to Kang, its primary importance was to serve as a sedimentation basin, as storage and water regulator, and as a reservoir for dry seasons. Steinkeller (1988) reports the sizes of *naĝkud*s, mentioned in several tablets, ranging from 12 m length, 6 m width and 3 m height up to 72 m length, 12 m width and 5 m height. The design of the *ka*, (𒅗, literally "mouth" in Sumerian, indicated by 3 in Fig. 2.14), should be in such a way to avoid erosion. A characteristic of *naĝkud* is to have a width much smaller than its length (a tablet records one 1 m width and 36 m length). It has been argued that they are too narrow to be efficient as storing facilities, being its primary function as an element to distribute water (Steinkeller, 1988, Pemberton et al., 1988). Efficiency of *naĝkud*s as water storage facilities is highly reduced due to evaporation (Pemberton et al., 1988). *naĝkud* as a silt retention device has also been questioned because 'there is a widespread view amongst farmers that silt laden water is preferable to clear water for irrigation…', and its function as a regulator is favored for some authors (Pemberton et al., 1988). Although *naĝkud*s are mentioned in pre-Sargonic Lagash, (ca. 2500–2340 B.C.) and they were an essential part of the canal system in the Ur III period (2112–2004 B.C.), they are not mentioned in later tablets, being replaced by large reservoirs, dams or swamps in the old Akkadian (2500–1950 B.C.) and old Babylonian (1950–1530 B.C.) (Kang, 1973, Steinkeller, 1988). The downstream end or outlet of a canal is called *kun* (𒆲, literally "tail" in Sumerian). Under this word, reservoir or storage basin is also understood (Halloran, 2006; Horowitz, 1998, p. 85).

To raise the water level in the main canal, (, *id* in Fig. 2.14), barriers were built, as in Fig. 2.14, number 5. They were called *kun-zi-da* , also ^{giš}*kešda* or ^{giš}*kešra* , and have been translated as dam, weir, sluice, floodgate (Halloran, 2006). Dams were constructed with fire-baked clay bricks and earth (Walters, 1970; Powell, 1988). Cooper (2002) mentions a brick inscription of king Enmetena of Lagash that says that 648,000 baked bricks and 1,849 gur (about 265 m^3) of bitumen were used in a dam in the Lumagina-du Canal. An exception in which stone carefully cut was used as building material instead of baked bricks is a small dam from the middle of the second millennium B.C., in nahr ed-Delbe (Tell Rash Shamra, in current Syria), (Calvet, 1990). Dimensions are rather modest: length 7 m, 3 m wide, and 1.60 m height. In mathematical texts, dams are presented as rectangular prisms. There is a text giving the following aspect ratio height:thickness:length = 1:3.333:13.333 (Powell, 1988).

Regarding the canals, the principal, *id*, could be 120 m width or more (Nemet-Nejat, 1993, p. 45), large enough to permit navigation. Frequently, the *id*s have levees or dikes (, *eg*) with small canals (, *par*) running on top of them. The dimensions of one of these small canals are given in an administrative tablet from UrIII period: 198 m length, 1 m width and 0.25 m depth (Waetzold, 1990). The principal canal, *id*, feeds smaller ones, which can feed others. Two texts regarding the irrigation system of Umma, mention that the branch canal depth is 0.5–1 m, one of them having 6 m width with a length reaching up to 1,710 m (Waetzold, 1990). The secondary canals can be as narrow as 1.00–1.25 m in width and 0.50–2.25 m in depth (Nemet-Nejat, 1993). The material from the excavation was probably used to raise the levees, increasing the canal depth. Although most of the mathematical exercises deal with rectangular canals, probably it is a simplification of trapezoidal shaped, in order to facilitate computations. Two trapezoidal channels are presented in Neugebauer and Sachs (1986, pp. 80–81), where the concept of side slope is introduced, measured as the horizontal distance per 1 unit of length (1 *kùš*[11]) in the vertical. Side inclination in both canals is V:H = 1:0.5. At what extent exercise tablets deal with real canal projects is hard to say, but probably they were basically examples used to train scribes. However, dimensions 'probably correspond more or less to those of real irrigation ditches' (Powell, 1988).

Maintenance of canals was a continuous task. Major canals were supervised by high officials who reported directly to the king. Large gangs of workers were necessary to have the canals free of silt, demanding the removal of enormous amounts of mud. For example a letter from Sin-iddinam to Hammurabi mentions 90 *iku* of soil to be cleared, corresponding to about 1,800 m^3 or 600 man-days of work (Neuman, 1980). Workers involved men from the localities situated on the banks of the canal including soldiers. Lesser canals were supervised by the local authority (Renger, 1990).

Although irrigation systems share common elements in all Mesopotamia, the canal layout depended on the particular physical landscape of the region to be

---

[11] 1 *kùš* = 0.5 m.

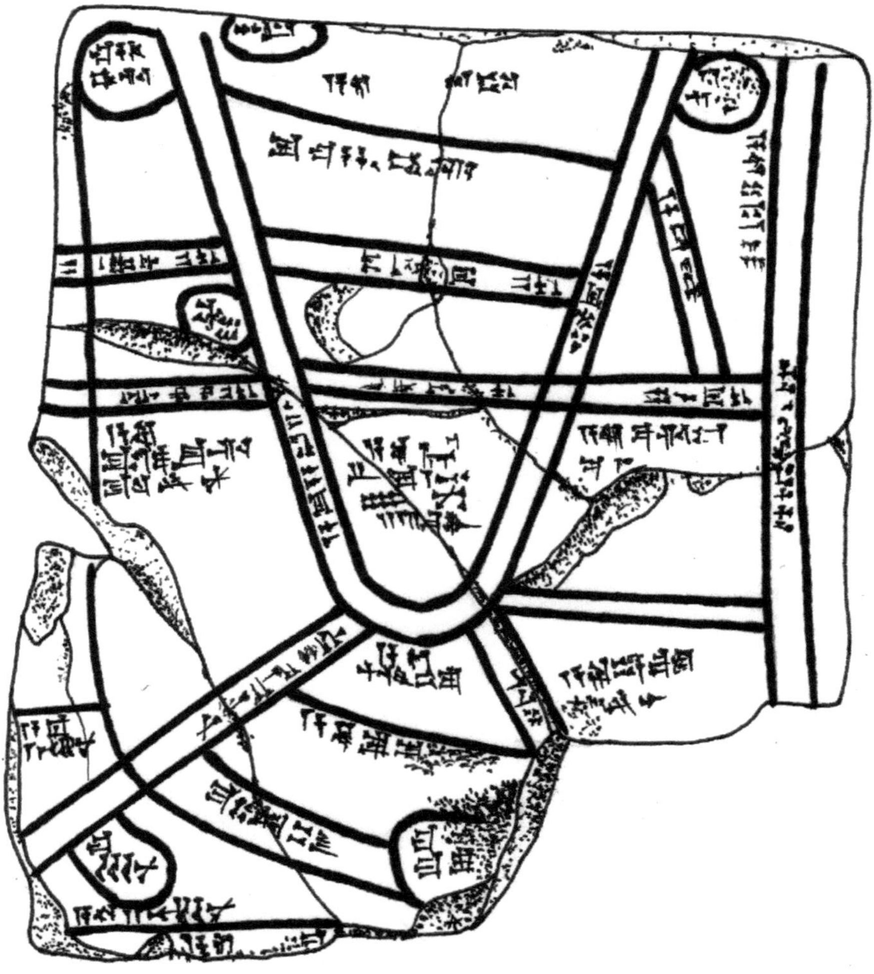

Fig. 2.15 Map showing the canals and irrigation fields near Nippur (ca. 1500 B.C.) Line drawing by the author from the tablet CBS 13885 in the University Museum, University of Pennsylvania

irrigated. Figure 2.15 shows a tablet depicting a plan of the fields of Nippur, ca. 1500 B.C. Canals are easily distinguished in the map. Postgate (1994) and Buccellati (1990) have proposed a layout of agricultural cells in South Mesopotamia (Fig. 2.16) and middle Euphrates, Syria (Fig. 2.17), respectively. In each case, the irrigation system is adapted to the local topography. In the first case, the salty water table restricts the cultivable area located in the lower land. Levees were stabilized by planting large vegetation on their banks. Figure 2.17 is adapted from Buccellati's interpretation of the fields of city of Terqa whose inhabitants took advantage of the peculiarities of the terrain to build a reservoir, erecting a dam in a depression that collected water from a wadi to be used to irrigate by bucketing nearby fields.

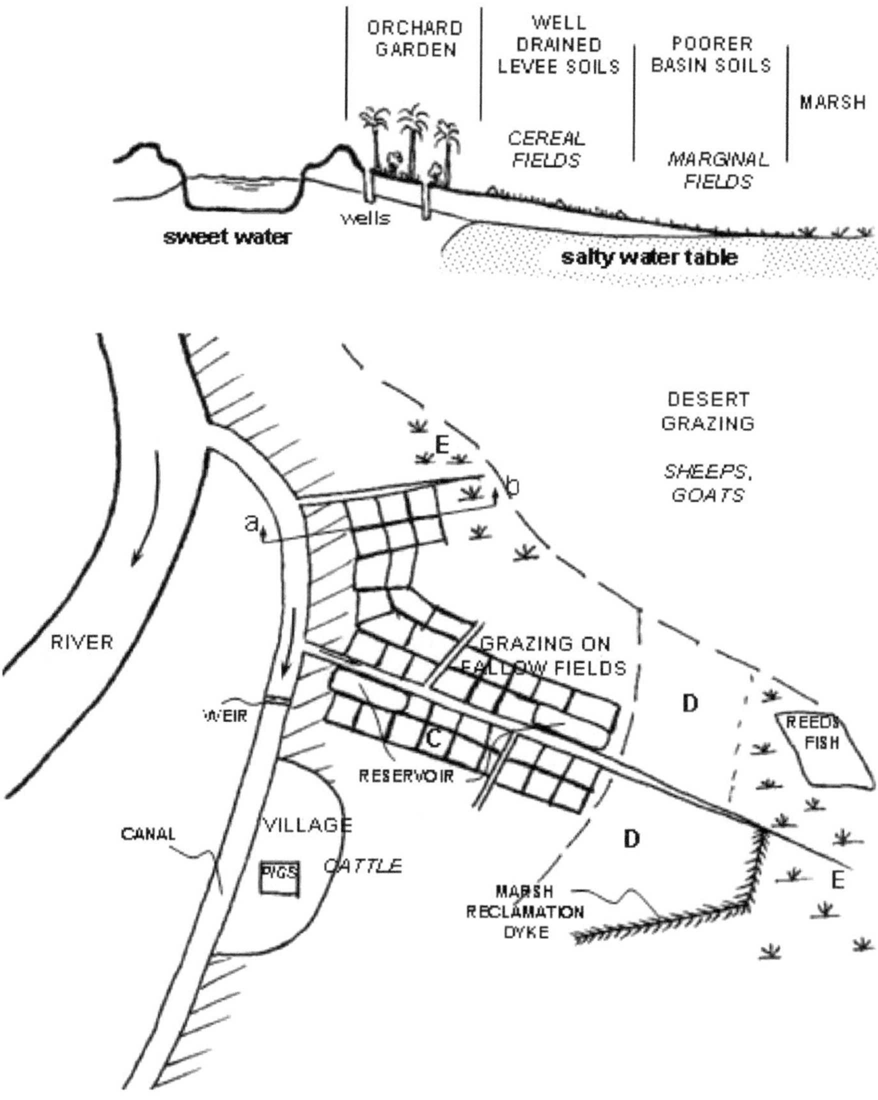

Fig. 2.16 Hypothetical layout of an agricultural cell in South Mesopotamia. (Adapted from Postgate, 1994, p. 175, Fig. 9.1)

2.5 Silting and Salinization

The irrigation techniques developed by Mesopotamians were magnificent, allowing them to increase agriculture production, maximizing the land cultivated. However, overexploitation of land and water resources for agriculture affected the environment. Canal branches and flood protection structures built to reduce or avoid flood

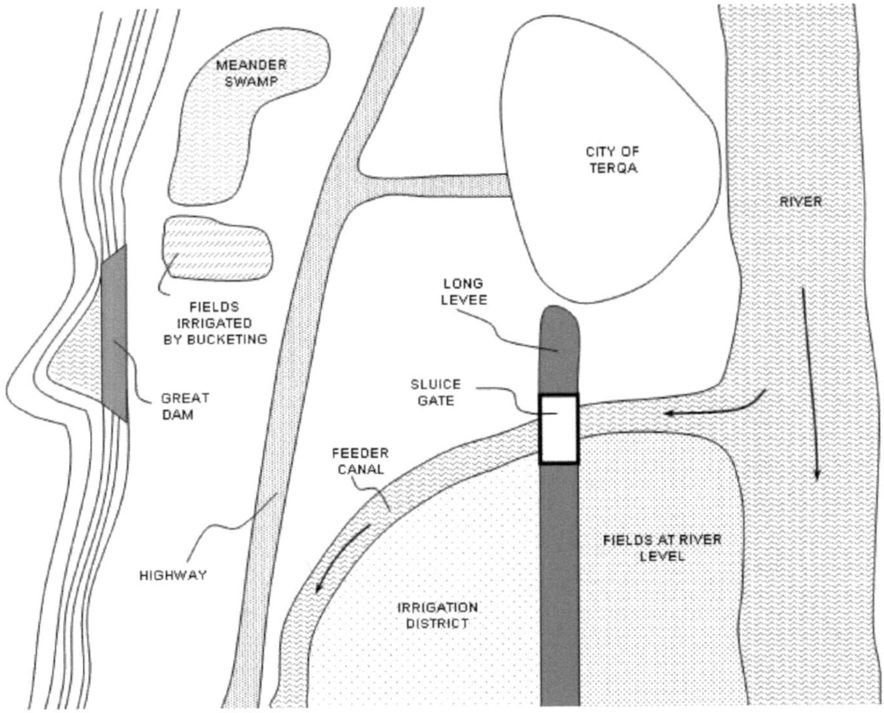

Fig. 2.17 Schematic layout of an agricultural complex in middle Euphrates. (Adapted from Buccellati, 1990, p. 163, Fig. 1)

surges intensified silt deposition. Also, over-irrigation brought salts into the surface soil, becoming infertile. The above problems affected settlement patterns, land-use and sociopolitical control (Walters 1970; Jacobsen and Adams, 1958) and they are reflected in the literature, as mentioned in Tablet II of Atra-Hasis epic: '. . .the black fields became white, / the broad plain was choked with salt. . .' (Lambert and Millard, 1999).

Given the topographic and climatic features of Mesopotamia, part of the salt is absorbed by colloidal clay particles and part is washed down into the water table, which has a very limited capability of moving them away, facilitating salt accumulation in the water table. According to Jacobsen and Adams (1958), "ancient control of the water table was based only on avoidance of overirrigation and on the practice of weed-fallow in alternate years".

Existing records establish that three major events associated with salinization have taken place in antiquity in a time span of almost four millennia, from 2400 B.C. to 1200 A.D. (Jacobsen and Adams, 1958). The gravest was the first, which happened in southern Mesopotamia between 2400 and 1700 B.C., and originated from conflict related to the use and control of water in irrigation channels fed by the Euphrates between Umma and Girsu. Both cities had been involved in a long dispute over a fertile border district watered by the Euphrates. Umma, located upstream controlled the water supply to the Lagash region (whose capital was Girsu). At some point of the conflict, one of the Girsu's rulers decided to build a large canal to bring water from the Tigris in order to be independent of the limited water delivered from the Euphrates. Although the canal could supply enough water to the region without interference from Umma, it increased seepage and flooding in addition to overirrigation, favouring conditions for soil salinization. Jacobsen and Adams (1958) stated that 'by 1700 B.C. this canal had become large and important enough to be called simply "the Tigris", and it was supplying a large region west of Girsu that formerly had been watered only by the Euphrates'.

The resulting salinization due to such a large canalization projects, including dams and other related facilities are present until nowadays, covering the complete region.

To handle soil salinization people replaced wheat by more salt tolerant crops. Thus, Jacobsen and Adams (1958) indicate that about 3500 B.C., the same proportion of wheat and barley were cultivated in southern Mesopotamia. Because barley is more salt tolerant, one thousand years later, it accounted for 83% of the crop. About 2100 B.C., barley accounted for more than 98%, and by 1799 B.C., the cultivation of wheat 'had been abandoned completely in the southern part of the alluvium'. The same decline of barley yield took place. At about 2400 B.C., the average yield was 2537 lt/ha. In 2100 B.C., it had diminished to 1460, and by about 1700 B.C., 'had shrunk to an average of only 897 lt/ha'.

2.6 Conclusion

The water resource provided by Euphrates and Tigris rivers shaped the society of the people living in Greater Mesopotamia. All aspects of their life were, in some way or another, conditioned by these two rivers. No doubt that they favored the flourishing of a civilization that reached high levels of development. Water was present from the very beginning; it was the *prima matter* in Mesopotamian cosmogony. It was the source of food, but also it can go out of control as floods. In order to take advantage of the resource, complex networks of canals were developed. They demanded elaborated waterworks, and qualified personnel to operate them. But knowledge of what we could call "applied hydraulics" was not limited to irrigation systems. Not less remarkable is the high level reached in the urban development, in particular "sanitary engineering". We wonder why it was lost.

However, overdevelopment of Mesopotamian irrigation was not sustainable in the long term. To satisfy the increasing demand on agricultural production has not been free, and since man started altering his environment he has had to pay a cost: silting and salinization.

Acknowledgments The author acknowledges permission to publish or reprint the following figures: Figure 2.2 Dr. Bron Lipkin, London, UK. Figure 2.5, Ashmolean Museum, University of Oxford. Figure 2.6, Field Museum Press. Figure 2.14, University of Illinois Press. From *Sumerian and Akkadian Cuneiform Texts in the Collection of the World Heritage Museum of the University* *of Illinois*. Copyright 1973 by the Board of Trustees of the University of Illinois. Figure 2.16, Routledge, Taylor and Francis Group.

The author also acknowledges Ms. Mirella Biancardi, Librarian of the Department of Civil Engineering of the University of Chile, who helped to get some of the references here cited.

References

Adams RMcC (1974) Historic patterns of Mesopotamian irrigation agriculture. In: Irrigation's impact on society, Anthropological papers No. 25, The University of Arizona Press, Tucson, Arizona, pp. 1–6

Alizadeh A (1996) Chogha Mish. The first five seasons of excavations, 1961–1971. Oriental Institute Publications, vol. 101, The University of Chicago Press, Chicago, IL

Alizadeh A (2008) Chogha Mish II. The development of a prehistoric regional center in lowland Susiana, southwestern Iran. Final Report on the last six seasons of excavations, 1972–1978. Oriental Institute Publications, vol. 130, The University of Chicago Press, Chicago, IL

Boehmer RM (1965) Die Entwicklung der Glyptik während der Akkad-Zeit., de Gruyter, Berlin, Tafel LX, seal 716.

Bonneterre D (2003) Gestion des crises et droits de l'eau en Mésopotamie à l'époque de Hammurabi http://www.er.uqam.ca/nobel/c2203/Doc..pdf

Breasted JH (1916) Ancient times. A history of the early world, Ginn and Company, Boston, p. 101

British Museum (1909) Cuneiform Texts from Babylonian Tablets, &c., in the British Museum, Part XXVI. Harrison and sons, London

Buccellati G (1990) "The rural landscape of the ancient zor: the Terqa evidence", in B. Geyer (Ed.), Techniques et practiques hydro-agricoles traditionelles en domaine irrigué. Actes du Colloque de Damas. Tome 1. Librairie Orientaliste Paul Geuthner, Paris, pp. 155–169

Calvet Y (1990) Un barrage antique à Ras Shamra. In Geyer B (ed) Techniques et practiques hydro-agricoles traditionelles en domaine irrigué. Actes du Colloque de Damas. Tome 2. Libraire Orientaliste Paul Geuthner, Paris, pp. 487–499

Cooper J (2002) Reconstructing history from ancient inscriptions: The Lagash-Umma border contlict. Revised third printing. Undena, Malibu

Dalley S (1993) Ancient Mesopotamian gardens and the identification of the hanging gardens of Babylon resolved. Garden History 21(1):1–13

Dalley S, Oleson JP (2003) Sennacherib, Archimedes, and the water screw. Technol Culture 44(1):1–26

Delougaz P, Hill HD, Lloyd S (1967) Private houses and graves in the Diyala region. Oriental Institute Publications, vol. LXXXVIII, The University of Chicago Press, Chicago, IL

Durand (1990) Problèmes d'eau et d'irrigation au royaume de Mari. In Geyer B (ed) Techniques et Practiques hydro-agricoles traditionelles en domaine irrigué, Actes du Colloque de Damas. Tome 1. Libraire Orientaliste Paul Geuthner, Paris, pp. 101–142

Finkelstein JJ (1966) The Laws of Ur-Nammu. J Cuneiform Studies 22:66–82

Gibson McG (1974) Violation of fallow and engineered disaster in Mesopotamian civilization. In: Irrigation's impact on society, Anthropological papers No. 25, The University of Arizona Press, Tucson, AZ, pp. 7–19

Glassner JJ (1993) Chroniques Mésopotamiennes, Les Belles Lettres, Paris

Halloran JA (2006) Sumerian lexicon. Logogram Publishing, Los Angeles

Harper RF (1904) The Code of Hammurabi, king of Babylon, 2nd edn. The University of Chicago Press

Horowitz W (1998) Mesopotamian Cosmic Geography, Eisenbrauns, Winona Lake, IN

Jacobsen T (1946) Sumerian mythology: a review article. J Near Eastern Studies 5(2):128–152

Jacobsen T (1960) The waters of Ur. Iraq 22:74–185

Jacobsen T, Adams RM (1958) Salt and silt in ancient mesopotamian agriculture. Science 126(3334):1251–1258

Jacobsen T, Lloyd S (1935) Sennacherib's aqueduct at Jerwan, Oriental Institute Publications, vol. XXIV, The University of Chicago Press, Chicago, IL
Kang ST (1973) Sumerian and Akkadian texts in the collection of the world heritage museum, vol. 2, Universtiy of Illinois Press,429 pp
King LW (1902) The Seven Tablets of Creation, vol. I. Luzac's Semitic text and translation series, vol. xii–xiii, Luzac and Co., London
King LW (1918) Legends of Babylon and Egypt in relation to Hebrew tradition, Oxford University Press, London
Lambert (1996) Babylonian Wisdom Literature, Eisenbrauns, Winona Lake, IN
Lambert WG (2007) Babylonian Oracle Questions, Eisenbrauns, Winona Lake, IN
Lambert WG, Millard AR (1999) Atra-Hasis, the Babylonian Story of the Flood, Eisenbrauns, Winona Lake, IN
Lara Peinado F (1996) Himno al Templo Eninnu. Cilindros A y B de Gudea, Editorial Trotta, Madrid
Layard AH (1851) Popular account of discoveries at Nineveh, John Murray, London
Layard AH (1882) Nineveh and Babylon. A narrative of a second expedition to Assyria during the years 1849, 1850, & 1851. John Murray, London
Ludwig W (1977) Mass, sitte und Technick des Bauens in Habuba Kabira Sud. In:Margueron JC (ed) Le Moyen Euphrate, zone de contacts et d'échanges, Actes du Colloque deStrabourg, pp. 63–74
Martin RA (1940) Ancient Seals of the Near East, Anthropology Leaflet 34, Field Museum of Natural History, Chicago
Nemet-Nejat R (1993) Cuneiform Mathematical Texts as a Reflection of Everyday Life in Mesopotamia. American Oriental Society, New Haven, CT
Neugebauer O, Sachs A (1986) Mathematical Cuneiform Texts. American Oriental Society, New Haven, CT
Neuman J (1980) Five letters from and to Hammurapi, king of Babylon (1792–1750 B.C.), on water works and irrigation. J Hydrol 47:393–397
The Oriental Institute (2004) The Assyrian Dictionary, vol. 7, I and J, Fifth Printing, Chicago, IL
Pemberton W, Postgate JN, Smyth RF (1988) Canals and bunds, ancient and modern. Bull Sumerian Agriculture, IV:207–221
Postgate JN (1994) Early Mesopotamia. Society and economy at the dawn of history. Routledge, New York
Powell MA (1988) Evidence for agriculture and waterworks in Babylonian mathematical texts. Bull Sumerian Agriculture IV:161–172
von Reber F (1882) History of Ancient Art. Harper & Brothers, New York
Renger J (1990) Rivers, water courses and irrigation ditches and other matters concerning irrigation based on Old Babylonian sources. Bull Sumerian Agriculture V:31–46
Roux G (1992) Ancient Iraq, 3rd edn. Penguin Books, England
Schmandt-Besserat D (1992) Before Writing, vol. I, From counting to cuneiform. University of Texas Print, Austin.
Steinkeller P (1988) Notes on the irrigation system in third millennium southern Babylonia. Bull Sumerian Agriculture IV:73–92
Stordeur D (2000) L'espace villageois d'El Kowm 2-Caracol. Organisation, limites et traditions architecturales. In Stordeur D (ed) El Kowm 2 : une île dans le désert – La fin du néolithique précéramique dans la steppe syrienne.CNRS Editions, Paris, pp. 87–94
Vallet R (1997) Habuba Kebira ou la naissance de l'urbanisme. Paleorient 22/2:45–76
Waetzold H (1990) Zu den Bewässerungseinrichtungen in der Provinz Umma. Bull Sumerian Agriculture V:1–29
Walters SD (1970) Water for Larsa. Yale University Press, New Haven, CT
Wittfogel KA (1957) Oriental despotism : a comparative study of total power. Yale University Press, New Haven, CT

Chapter 3
Water Technology in Ancient Egypt

Larry W. Mays

3.1 Introduction: The Nile River

Throughout history humans have been fascinated with the Nile River, especially the Egyptian part of the Nile. The birth of this great civilization has been traced back to a time between 11,000 and 10,000 years ago. Around five thousand years ago this civilization started depending entirely on the Nile River and its annual inundation. This chapter traces the history of water engineering in ancient Egypt starting with the uses of water from the annual inundation of the Nile River for natural irrigation in the Predynastic period to the development of methodologies to advance the use of the Nile River for irrigation.

For thousands of years the people of Egypt have owed their very existence to a river that flowed mysteriously and inexplicably out of the greatest and most forbidding desert in the world (Hillel, 1994). Herodotus (Book II.5, 440 B.C.) in Book II.5 stated

> That they said of their country seemed to me very reasonable. For any one who sees Egypt, without having heard a word about it before, must perceive, if he has only common powers of observation, that the Egypt to which the Greeks go in their ships is an acquired country, the gift of the river. The same is true of the land above the lake, to the distance of three days′ voyage, concerning which the Egyptians say nothing, but which exactly the same kind of country.

The ancient Egyptians depended upon the Nile not only for their livelihoods, but they also considered the Nile to be a deific force of the universe, to be respected and honored if they wanted it to treat them favorably. Its annual rise and fall were likened to the rise and fall of the sun, each cycle equally important to their lives, though both remaining a mystery. Since the Nile sources were unknown up until the 19th century, the Ancient Egyptians believed it to be a part of the great celestial ocean, or the sea that surrounds the whole world.

L.W. Mays (✉)
School of Sustainable Engineering and the Built Environment, Arizona State University, Tempe, AZ, 85287-5306, USA
e-mail: mays@asu.edu

L.W. Mays (ed.), *Ancient Water Technologies*, DOI 10.1007/978-90-481-8632-7_3,

Fig. 3.1 View of the Nile River in Egypt (courtesy of NASA)

The Nile River (shown in Fig. 3.1), nearly 6,650 km in total length, draining an estimated 3,350,000 km^2, which is about one-tenth of the African continent with catchments in nine different countries. In contrast the Amazon River has a length of 6,400 km with an estimated drainage area of 7,050,000 km^2. There has been disagreement as to which of these two rivers is the longest. The annual discharge of

Table 3.1 Chronology of Ancient Egypt (after Butzer, 1976 and Strouhal, 1992)

Epi-Paleolithic Period	5000–4000 B.C.
Predynastic Period	5200–3050 B.C.
Early Dynastic Period	3050–2700 B.C.
Dynastic Period	2700–332 B.C.
Old Kingdom (Pyramid Age)	2700–2215 B.C.
First Intermediate, 7th–10th dynasties	2250–2040 B.C.
Middle Kingdom, 11th–12th dynasties	2040–1715 B.C.
Second Intermediate, 13th–17th dynasties	1715–1570 B.C.
New Kingdom, 18th–20th dynasties	1570–1070 B.C.
Late Period, 21st–31st dynasties	1070–332 B.C.
Graeco-Roman Period	332 B.C.–641 A.D.
Macedonian	332–323 B.C.
Ptolemaic	323–30 B.C.
Roman – Byzantine	30 B.C.–641 A.D.
Arab Conquest	A.D. 641

the Nile is only 84×10^9 m^3 as compared to $5{,}518 \times 10^9$ m^3 for the Amazon River. The Nile Delta, which is also known as Little Egypt, is a giant triangle of land, as shown in Fig. 3.1, and is approximately 200 km wide along the Mediterranean Sea with the apex at Cairo, about 160 km inland. To the south of the delta apex and to the west of the Nile is the Faiyum Depression, discussed later. Main sources of the present-day Nile are the Sudan basin and the Ethiopian Highlands. Being one of the most predictable rivers in the world, the Nile flood is seldom sudden or abrupt and is timely, in contrast to the floods of the Tigris and the Euphrates which have more abrupt floods. Average duration of a flood was about 110 days. The beginning of the rise of the Nile begins in June, with the maximum rise of the river usually occurring in the later part of September and the early part of October.

Throughout the discussions in this chapter references are made to various time periods associated with Egyptian history. A chronological framework for ancient Egypt is given in Table 3.1.

3.2 Water Engineering: Predynastic

First actual recorded evidence of water management was the mace head (Fig. 3.2) of King Scorpion, the last of the Predynastic kings, which has been interpreted as a ceremonial start to breaching the first dyke to allow water to inundate the fields or the ceremonial opening of a new canal (Strouhal, 1992). Similarly others have interpreted the main part of the mace-head of the king as depicting irrigation work under his supervision. This mace-head indicates that the ancient Egyptians began practicing some form of water management for agriculture about 5,000 years ago.

One of the key unknowns in Egyptian history is the time when people began artificial irrigation, in particular canal systems, consciously as a means to improve the natural effect of the Nile (Strouhal, 1992). Canals allowed the flow of floodwater to locations that could not be reached otherwise, and when the Nile flood levels were

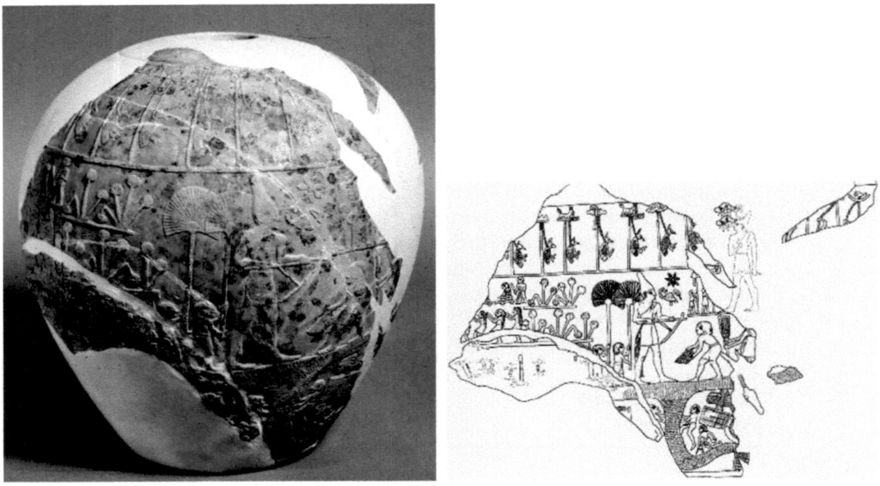

(a) (b)

Fig. 3.2 Mace-head of the King Scorpion – The first recorded evidence of water management. (**a**) Photo of Mace-head of King Scorpian taken at the Ashmolean Museum in Oxford by Henning Fahlbusch (photo courtesy of Henning Fahlbusch). Lower part of the mace head shows a waterway that bifurcates to two channels (irrigation canals). The king is holding a large hoe while attendants are ready with a fiber basket and broom. Two other workers, with hoes are working on the lower canal. This document has been interpreted as leaving little doubt that the transition from natural to a modified and ultimately artificial irrigation had been completed by the end of the Predynastic period. (**b**) Reconstruction of Mace-head of King Scorpion (used with permission of Krzysztof M. Cialowicz)

low, canal networks made artificial watering easier. Canals also were built for water traffic and for the drainage of marshes. The shift from natural to artificial flood irrigation was accomplished by the late Predynastic times. As long as the annual Nile floods were persistently good, the Predynastic population density was not large enough to warrant artificial irrigation. The average Nile flood would allow a single crop season over possible two-thirds of the alluvial surface.

3.3 Water Engineering: Dynastic

3.3.1 Artificial Basin Irrigation

Artificial irrigation was established by the 1st Dynasty (Butzer, 1976). This included deliberate flooding and draining using sluice gates and water contained by longitudinal and transverse dikes. Artificial irrigation increased the area of annual cropland in relation to the flood stage; retained water in the basin after smaller floods; and allowed second and even third crops in some basins. This form of water management, called basin irrigation, consisted of a network of earthen banks, some parallel

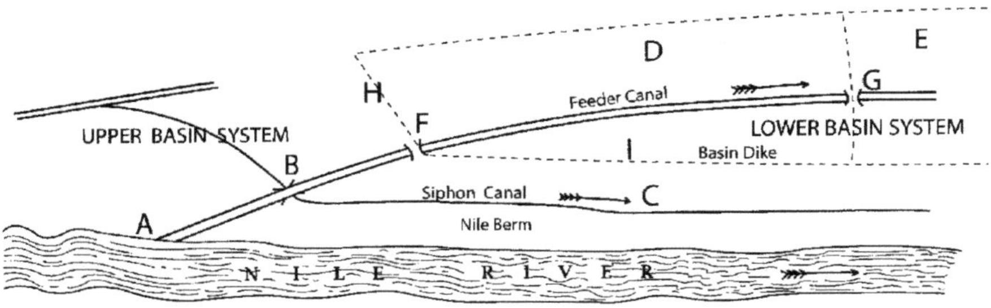

Fig. 3.3 Plan of basin irrigation. (**A**) head of basin canal, (**B**) siphon of siphon canal, (**C**) High land under sorghum, (**D, E**) basins, (**F, G**) regulators, (**H**) transverse dike, (**I**) longitudinal dike (modified after Willcocks and Craig, 1913 and Said, 1993)

to the river and some perpendicular to the river that formed basins of various sizes. Floodwaters were diverted into the basins where the water was allowed to saturate the soil with the remaining water drained off to a down-gradient basin or to a canal. After the draining process was completed in a basin, crops were planted. King Menes, the founder of the first dynasty in 3100 B.C. traditionally has been known as the first to develop a major basin irrigation project. Basin irrigation was carried out on a local scale as opposed to being centrally managed on a national scale.

Artificial basin irrigation was based upon the inundation of the Nile floodplain starting in early August. The floodplain was divided into basins ranging in size from 2000 feddans (1 feddan = 4,200 m^2) in the upper part of Egypt to 20,000 feddans in the Nile delta (Said, 1993). Figure 3.3 illustrates the concept of basin irrigation in which the basins were supplied with water by feeder canals. The bed level of the feeder canal was midway between low Nile and ground level with a natural downstream slope less than the slope of the Nile. Each canal supplied water to an average of eight basins arranged in succession. Dikes (levees) separated the basins with controls (masonry regulators) on the earthen embankments to control the flow of water into the basin. Average depths of water in the basins varied according to the local flood volume and stayed in the basins between 40–60 days after which the basins were drained (Said, 1993). The basins were very level as a result of the water laden alluvium that deposited in the basins. During years of low flow in the Nile basins were drained into the next downstream basin instead of back to the river.

3.3.2 *Lift Irrigation*

Improvement of irrigation technology continued through the Dynastic period. The shift to lift irrigation was well under way during the 18th Dynasty and was effective

Fig. 3.4 Present day use of the shaduf (photo courtesy of Henning Fahlbusch)

by Roman times (Butzer, 1976). Sometime after 1500 B.C. the ancient Egyptians began lift irrigation with the shaduf (shadouf), already in use in Mesopotamia, is shown in Figs 1.1 and 3.4 for irrigating small plots. This device allowed the irrigation of crops near river banks and canals during the summer. The shaduf had a bucket and rope attached to the one end of a wooded arm with a counter balance at the other end of the arm. This device typically lifted water up to 1.5 m. One shaduf could irrigate approximately 0.3 acres of land in 12 hours.

3.3.3 Sadd-el-Kafara Dam

The Sadd-el-Kafara dam (Dam of the Pagans) was constructed about 2600–2700 B.C. (Garbrecht, 1985). Professor Henning Fahlbusch (personal communication) has confirmed that the dam was constructed around 2650 from his dating studies. This dam was the first attempt at storing water on a large scale (Murray, 1955; Garbrecht, 1985). Possibly older dams include the Jawa reservoir in Jordan and diversion dams on the Kasakh River in the southern part of the former Soviet Union. However these structures were much smaller than the Sadd-el-Kafara dam allowing us to refer to this dam as the world's oldest large-scale dam (Garbrecht, 1985, Schnitter, 1994). This dam was constructed in the Wadi Garawi, seven miles southeast of Helwan (also Heluan) and approximately 30 km south of Cairo on the eastern Nile bank for flood protection of installations in the lower wadi and in the Nile valley (see Fig. 3.5). This dam was still in the construction phase (about eight

(a) (b)

Fig. 3.5 Sadd el-Kafara – Dam of the Pagans in the Wadi Garawi. (**a**) View of northern wing of dam showing slope protection on the upstream face. (**b**) Close-up view showing the revetment blocks (photos courtesy of Henning Fahlbusch)

to ten years) when it failed as a result of a flood catastrophe. There was no channel or tunnel to divert the river around the construction site (Schnitter, 1994). It was another eight centuries before the Egyptians constructed another dam. Sadd el-Kafara dam was discovered in 1885 by the German archaeologist G. Schweinfurth (Smith, 1971).

Dimensions of the dam were 14 m in height and 113 m in crest length with a 0.5 million m^3 storage capacity (Schnitter, 1994). The dam consists of two rock fill sections with the space between them filled with a core of silty sand and gravel (Garbrecht, 1985). The volume of the fill was 87,000 m^3. Revetment (limestone) blocks (around 17,000 according to Schnitter, 1994), of about 50 lbs each arranged in the form of stairs 28 cm high, were used to cover the surface of the rock fill (Murray, 1955). The facing of the dam with the revetment blocks is shown in Fig. 3.5. Smith (1971) estimates the catchment area above the dam to be 186 km^2. According to Garbrecht (1985) by modern standards the dam is stable and would have withstood the floods if it had been completed as opposed to having been in the construction phase. The design of the dam basically was correct and was protected from water damage that would have been caused by water bypassing the dam on the left bank or overtopping.

A sag in the structure existed along the top of the dam that diminished the effective height of the dam. The maximum amount of the breach cannot be determined because of the extent of the breach, 44 m wide, and the sag probably did not occur after the breach because there is no sign of slip in the abutments (Murray, 1955). Possibly the sag was caused by general settlement in the loosely compacted structure and being constructed, as no mortar was used in the dam. Ancient Egyptians did not use mortar as a cementing material but used it as a lubricant in moving heavy blocks and for purposes of leveling.

No spillways were provided for the dam indicating that the reservoir was not built for the purposes of irrigation. This is even more evident in the fact that there is no land capable of being cultivated in the Wadi el-Garawi. Because there was no spillway, most likely the reservoir was not to be completely filled. Murray (1955) believed that they never intended to fill the reservoir but merely intended to contain the largest flood for the Wadi el-Garawi. However as Fahlbusch (2004b) points out, it is assumed that the dam was constructed as a flood protection measure, but what was it protecting? In other words the purpose of the dam is still controversial. Fahlbusch (2004b) gives us something to think about, "It seems to be fact that the Sadd el Kafara dam was destroyed before its completion. But was this dam, with its extremely large dimensions really the first of it kind in Egypt?" He also states, "the courage and daring of the master-builders are still admirable even after more than 4000 years." Other excellent readings on the Sadd el Kafara dam include Fahlbusch (1996, 2004a).

3.3.4 Faiyum Depression

The Faiyum (or Fayum) Depression (Fig. 3.6), located about 60 km southwest of Cairo, is a huge (1,700 km^2), geological depression (below sea level) that begins 20 km west of the Nile Valley, extending into the Western Libyan desert region. A vast saltwater lake (Lake Moeris) was in the heart of this region until the Paleolithic Period. Historically, a natural channel, the Bahr Yusuf, branched off the Nile River about 334 km south of the Faiyum Depression and located along the valley's western escarpment and connected the Faiyum to the Nile River through the Hawara Channel. High water levels in the Nile resulted in the formation of a lake within the Faiyum. During the Old Kingdom a permanent lake existed in part of the depression. In the Middle Kingdom the kings directed that the Hawara Channel be cleared to permit excess flood waters from the Nile to enter the depression, sparing the Delta from flooding. After the flood the water drained from the Faiyum back to the Nile. Flood control was no longer deemed necessary by the time of the Ptolemaic kings (Graeco-Roman period) and the Faiyum was exploited for agriculture. The Bahr Yusef was used to convey irrigation water into the depression and then was dispersed by canals across the fields. Drainage water was conveyed to the deepest part of the depression to collect in the Lake Qarun. Prior to the time of the lowering of the lake, the Faiyum Depression was a natural storage for a large portion of the floodwaters that protected lands of Lower Egypt.

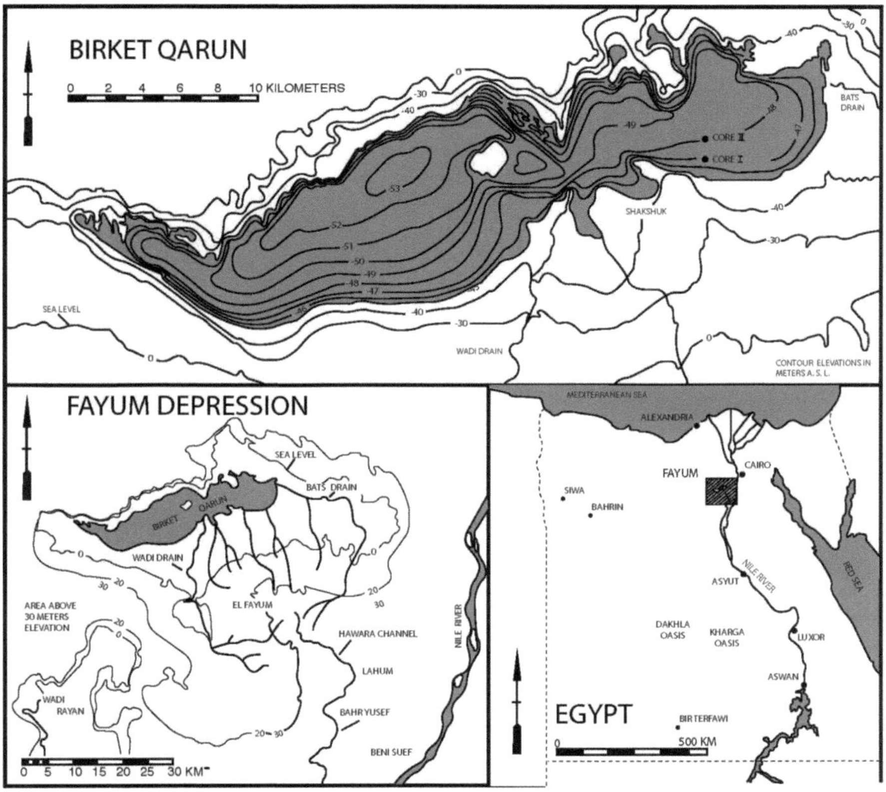

Fig. 3.6 Faiyum Depression Showing the Birket Qarun (Mehringer et al., 1979)

3.4 Water Management: Graeco-Roman Period

3.4.1 Irrigation

Alexander the Great, King of Macedon, conquered Egypt in 332 B.C., with little resistance from the Persians and was welcomed by the Egyptians as a deliverer. After Alexander's death in 323 B.C. Ptolemy ruled Egypt and in 305 B.C. took the title of King. As Ptolemy I he founded the Ptolemaic dynasty that ruled Egypt for nearly 300 years. Egypt became part of the Roman Empire as the province Aegyptus in 30 B.C. The Roman's interest in Egypt was the reliable delivery of grain to Rome. As a result Rome made no change to the Ptolemaic system of government. The Romans replaced Greeks in the highest offices but used Greeks to staff most of the administrative offices. Greek remained the language of government except at the highest levels. Unlike the Greeks, the Romans did not settle in Egypt in large numbers. Culture, education and civic life largely remained Greek throughout the Roman period.

Fig. 3.7 Archimedes screw (photo courtesy of Henning Fahlbusch)

(a) (b)

Fig. 3.8 Water wheels (photos courtesy of Henning Fahlbusch)

Irrigation of larger plots became possible using the Archimedes screw or tanbur (Fig. 3.7) and the waterwheel or saqiya (Fig. 3.8), which were introduced in the Ptolemaic times. The Archimedes screw consists of a water tight cylinder enclosing a chamber walled off by spiral divisions running from end to end. Water is lifted by turning the cylinder so that water is raised in successive divisions of the spiral chamber from the lowest portion first. Introduction of the saqiya (or waterwheel) during the early Ptolemaic times revolutionized lift irrigation. This device (Fig. 3.8) consists of a row of pots attached to the rim of a revolving wheel. The pots when dipped into an irrigation canal fill with water and are then lifted on the wheel to a height the diameter of the wheel (usually 3–6 m). The wheels were turned by oxens.

Lift irrigation coupled with the entrepreneurial system allowed Egyptian agriculture to expand and intensify during the Ptolemaic times. The waterwheel (saqiya) enabled the reclamation of the Fayum depression, which had formed the lake and allowed the storage of high floods since the Middle Kingdom. The spread of the saqiya during the Ptolemic and Roman times led to the introduction summer crops as well as flood crops. This in turn led to increased wealth of Egypt. The expansion was to a degree unmatched until a century ago, after the introduction of perennial irrigation a century ago (Butzer, 1976).

3.4.2 Reclamation of the Faiyum Depression

The reclamation project of the Faiyum Depression was performed by the Ptolemies in which Lake Moeris was dried up from a previous level of 20 m above sea level to about 2 m above sea level during the reign of Ptolemy I (323–285 B.C.). Figure 3.9 shows trends in Lake Moeris starting from around 5000 B.C. The dried up level of the lake has been surmised by Caton-Thompson and Gardner (1934) from the inferred level of a saqiya well northeast of Birket Qarun. The drying up of the lake during early Ptolemaic times could not have been caused from low flows in the Nile because the flows were adequately high during this time period (Said, 1993). To lower the lake Ptolemy II (285–246 B.C.) constructed an embankment near Lahun in order to control the flow of water from the Nile into the Hawara channel that flows to the depression. This embankment (dike today measures some 5,000 m in length and up to 4 m height, Schnitter, 1994) closed the gap between two hills with the exception of a single opening with a dam and a weir at Luhan. The weir was used to keep the level of the lake at 2 m above sea level. The canal system used to channel

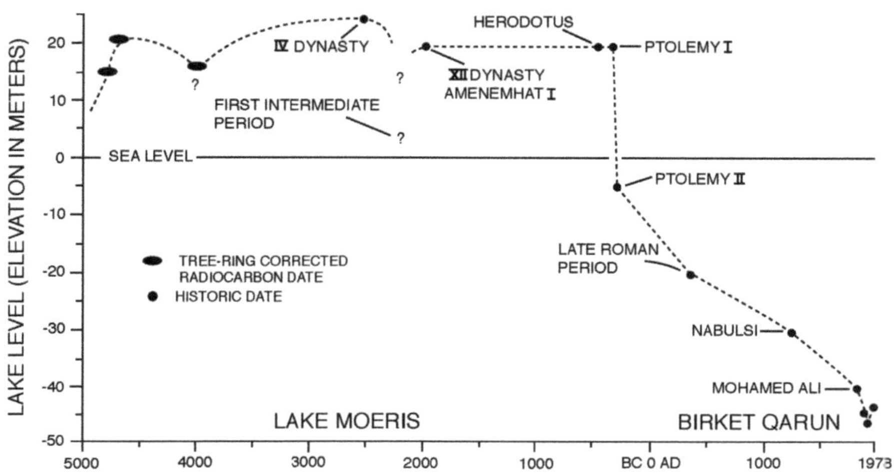

Fig. 3.9 Trends in Lake Moeris levels from 5000 B.C. (Mehringer et al., 1978)

water from the Nile River into the Faiyum Depression consisted of a radial network of relatively high gradient canals. This canal system was unique as compared to the canal systems used in the Nile valley and delta (Said, 1993). The reclamation project by the Ptolemaic engineers added approximately 325,000 acres of new and fertile arable land to Egypt (Said, 1993). This project along with the wide-spread use of the waterwheel significantly increased the wealth of Egypt and allowed the population to increase to an estimated 4.9 million people, the largest during the long history of Egypt prior to the nineteenth century.

3.5 Measurement and Long-Term Records of the Nile

Because the well-being of Egyptian society relied on the annual flows in the Nile River, the ancient Egyptians developed methodologies to measure and record flood levels. Oldest records of Nile flood levels were carved on a large stone monument during the Dynasty V (2480 B.C.). The Palermo Stone is the most valuable surviving fragment of the monument, named after the capital of Sicily were it is located in a museum. The Palermo Stone also records a number of early gods, that copper smelting was taking place, records forty ships that brought wood from and unknown region outside of Egypt.

Nilometers were used to measure the levels of the Nile River. At Karnak, Nile levels were marked on the quay walls of the great temple, dating from about 800 B.C. Three types of nilometers were used: simply marking water levels on the cliffs of river banks; utilizing flights of steps that led down to the river; and using conduits to bring river water to a well or cistern. At the Roda Nilometer, south of Cairo, there were 820 floods recorded between the 7th and 15th centuries, of which 73 percent were normal floods, 22 percent were low, and five percent were destructively high (Said, 1993).

The ancient lake sediments and shorelines in the Faiyum Depression (discussed later) in Middle Egypt also provide a record of Nile floods (Mehringer, et al., 1979; Hassan, 1986, 1998). Various lake levels are indicated by former shoreline features and deposits, which allow the inference of variations of the Nile flood discharge in prehistoric and Pharaonic times. The science of geometry arose from the need to perform new land measurements after every flood of the Nile.

3.6 Concluding Remarks

Fundamental interrelationships between humans and their environment in the Egyptian floodplain greatly influenced the evolution of land use patterns, the development of irrigation, and the spatial distribution of settlements. The Nile River valley is a seasonally inundated floodplain of the Nile, and being one of the most predictable rivers in the world, the Nile flood is seldom sudden or abrupt and is timely, in contrast to the floods of the Tigris and the Euphrates, which have more abrupt floods. Low Nile floods cause famines. Throughout history the advancements of irrigation of the Nile, starting from natural irrigation and advancing to artificial irrigation and then the development of lift irrigation with the shaduf and then the Archimedes screw (or tanbur), and the saqiya (or waterwheel). From a water management perspective, all evidence known suggests that flood control and irrigation, at the social and administrative levels, were managed locally by the rural population within a basin. The rise and sustainability of Egypt, with so many great achievements, was based primarily on the cultivating of grain on the Nile River floodplain, without a centralized management of irrigation. What is so unique is that Egypt probably survived for so long because production did not depend on a centralized state. Collapses of the government and changes of dynasties did not undermine irrigation and agricultural production on the local level. "The secret of Egyptian civilization was that it never lost sight of the past" Hassan (1998).

References

Butzer KW (1976) Early hydraulic civilization in Egypt: a study in cultural ecology. The University of Chicago Press, Chicago

Caton-Thompson G, Gardner EW (1934) The desert Fayum, Two volumes. Royal Anthropological Institute, London

Fahlbusch, H (1996) Ancient dams in Egypt. Transactions of the 16th Congress on Irrigation and Drainage, vol. 1G, New Delhi

Fahlbusch H (2004a) Men of dikes and canals. In: Hans-Dieter Bienert, Haser J (eds) Men of dikes and canals: The archaeology of water in the Middle East, pp. 1–16 (International Symposium held at Petra, Wadi Musa (H. K. of Jordan), June 15–20, 1999), Verlag Marie Leidorf GmbH, Rahden/Westf

Fahlbusch H (2004b) The Sadd el Kafara – The oldest high dam of the world. In: Hans-Dieter Bienert, Haser J (eds) Men of dikes and canals: The archaeology of water in the Middle East, pp. 365–368 (International Symposium held at Petra, Wadi Musa (H. K. of Jordan), June 15–20, 1999), Verlag Marie Leidorf GmbH, Rahden/Westf

Garbrecht G (1985) Sadd-el-Kafara: The world's oldest large dam. Water power & dam construction, pp. 71–76, July 1985

Hassan F A (1986)Holocene lakes and prehistoric settlements of the western Faiyum, Egypt. J Archaeological Sci 13:483–501

Hassan FA (1998) Climatic change, Nile floods, and civilization. Nature and resources, 32(2): 34–40

Herodotus (440 B.C.) The history of Herodotus, English version. Translated by G Rawlinson, http://classics.mit.edu/Herodotus/history.html

Hillel D (1994) Rivers of Eden: The struggle for water and the quest for peace in the Middle East. Oxford University Press, Oxford

Mehringer PJ Jr, Peterson KL, Hassan FA (1979) A pollen record from Birket Qarun and the recent history of the Faiyum Egypt. Quaternary Res 11:238–256

Murray GW (1955)Water from the desert: Some ancient Egyptian achievements. Geographical J 121:171–187

Said R (1993) The river Nile: Geology, hydrology and utilization. Pergamon Press, New York

Schnitter NJ (1994) A history of dams, the useful pyramids, A. A. Balkema, Rotterdam, Netherlands

Smith NA (1971) History of dams. Peter Davies, London

Strouhal E (1992) Life in ancient Egypt. Cambridge University Press

Willcocks W, Craig JI (1913) Egyptian irrigation. E. & F.N. Spon Ltd, London

Chapter 4
Ancient Greek Lavatories: Operation with Reused Water

Georgios P. Antoniou

4.1 Introduction

Lavatories can be an indicator of living standard and economic prosperity. Many remains of ancient lavatories have been found in the ancient Greek world, some dating back to the Minoan era. (i.e. Knossos' Palace). Many references to ancient lavatories have been recorded in numerous ancient Greek scripts. Despite the fact that many related archaeological finds are dated over a wide chronological range, the typical mature ancient Greek lavatory probably is from the Hellenistic period. This was the period of great evolution in Greek water technology. Chambers for that function are found not only in private houses but also in many public buildings and sanctuaries. (Gymnasia, Asclepieia, etc.)

The features of the typical ancient lavatory are the bench type seats with keyhole-shaped defecation openings and an underneath ditch. The ditch was both a water supply conduit for flushing and a sewer. The lavatory was usually situated in the area of the building most convenient for water supply and/or sewerage. In many cases the water for the flushing was reused either after other domestic or communal activities. Despite privacy lavatories were used in antiquity by many people simultaneously, from two to three people in the small domestic latrines and up to 60 people in the larger public latrines. Lavatories were used throughout the Roman Empire, with a more or less monumental appearance.

The subject of this chapter is derived mostly from experience through various architectural projects requiring bibliographic research. Installations for hygiene are one of the characteristic factors of living standard and economic prosperity. As a result the tourism industry evaluates accommodation facilities and other installations for tourists through inspection of their toilets.

G.P. Antoniou (✉)
Hellas, 44 Ioannou Soutsou, 11474 Athens, Greece
e-mail: antonioug@tee.gr

L.W. Mays (ed.), *Ancient Water Technologies*, DOI 10.1007/978-90-481-8632-7_4,

4.2 Terms – Etymology

Ancient Greeks used various terms such as άφοδος – afodos or απόπατος – apopatos but also αποχώρηση – apochorisi – withdrawal. The term θώκος – thokos refers to seat, throne, or chair. The last one is characteristic for the shape of the lavatories–as it will be presented analytically below- and for the equivalent portable or fixed utensils. These artifacts were used for defecation before the formation and predominance of the typical ancient lavatory. Its reference as σωτηρία – sotiria – salvation, for a public lavatory in ancient Smirni, is also rare. The existence of such installations was indeed salvation for the ancient city.

The term κόπρων – kopron probably referred to small cesspits. According to archaeological data there were no sewerage pipes which characterized the typical ancient lavatory, απόπατος – apopatos. Because of that lack of sewerage κοπροδόχοι – koprodochoi and κοπροθήκες – koprothikes were essential as well as κοπρολόγοι – koprologoi, those who gathered the sewage. Finally, it is evident that a specialized researcher identifies easily the ancient roots of most words that are used in modern Greece referring to defecation. It is also remarkable that the basic terms include the concepts of isolation and privacy that undeniably puts the prefix απο – apo.

4.3 Written Sources

4.3.1 Aristofanes

Aristophanes' comedies included the most significant ancient written reports on lavatories. Apart from the term απόπατος – *apopatos* that is mentioned in "Ploutos" (line 1185), in "Ekklisiazousai" (326, 351, 354) and in "Acharneis" (the Ambassador in the Dikaiopoli in row 81), its synonym άφοδος – *afodos* is also written (Ekklisiazousai 1059) as well as the relevant κόπρων – kopron in Thesmoforiazousai (line 485). Moreover in "Frogs" the sanitary paper of that era is reported as σπογγιά – sponghia, which was made of sponge (line 487). In this reference, Dionysus discusses with Xanthia sanitary paper. Finally, in "Peace" (l. 9), the κοπρολόγοι – *koprologoi* are characterized in depreciatory manner, as vituperation (Aristophanes et al., 1967).

4.3.2 Other Sources

In Demosthenes (25, 49), as well as in various inscriptions (i.e. IG2 1058II), Κοπρώνας– *Kopronas* is reported. In an inscription from Pergamon (OGI 483.220 in Athen. Mitt. 27, 1902) is reported the term αφεδρών– afedron out of αφοδεύω– *afodevo*, from which derives the contemporary αφοδευτήριο – afodeutiro. Ἄφοδος – *Afodos*. Apart from Aristophanes is reported also by Hippocrates (peri agmwn, 16) while in Epidemies (7, 47, 84) he used the term θώκο– *thoko*.

In the biography of Lykourgos, Ploutarhos refers to lavatory as αποχώρηση – *apochorisi I-withdrawal,* a familiar term after the contemporary word αποχωρητήριο – *apochoritirio* – lavatory. In that script is reported the problem of stench because of sewage in the streets of ancient Athens.

Polydeukis refers to *immovable lavatory (.. epi akinitou apopatou - on immovable apopatos)* to distinguish it from the existing vessels (Bethe E. 1900–37 10, 44). During the early classical period the ancient written sources quite often referred to odor in the streets of Athens and to regulations for preventing the throwing of wastes onto the streets!

4.4 Emergence of the Type and its Time Frame

Similar installations existed during the Minoan and the Mycenaean period (Angelakis et al., 2005, p. 212), but are not presented in this chapter. Even though latrines are mentioned in the scripts by the ancient writers, no public or private lavatories have been found that were dated during the Classical period. Researchers that dealt systematically with this subject, agree on this point (Neudecker, 1994). Cesspits in houses (κόπρων – *kopron*) constructed during the Classical period have been discovered north of Areios Pagos, by the American School of Classical Studies of Athens. However in Rhodes, small rectangular constructions under the roads just outside the houses definitely have been considered as koprons (Filimonos, 2000).

Moreover, clay containers for defecation (*κοπροδόχοι* – koprodochoi) are known (amides or skoramides from Athens), as well as earthen seats such as those at Olynthos, as illustrated in Fig. 4.1. These seats have a similar appearance as modern day ones. Because the seats did not have a bottom combined with the form of the lower edge, they were used either over cesspits or had some other mechanism for collection and drainage of excrements. Probably they were a pre-existing type of lavatory especially if we compare the shape of the apertures of the two types. The existence of such utensils in Olynthus (destroyed by Philipp II in 348 B.C.) could easily date them to the 5^{th}century B.C. In addition to that, in Olynthus was found an earthen utensil with a clay sewerage pipe. Its shape, according to the excavator, concludes that it was used along with something else that was not preserved, such as a wooden seat.

Finally recent discoveries in Epidaurus, specifically at the foundations of Avaton, is located probably one of the first stone toilet seats. After basic research into the time of appearance of lavatories with this mature layout, it probably appeared in the early 4th century B.C. Basic issues for this hypothesis are, first the absence of lavatories in the 5th century B.C. finds – however they are reported in the ancient scripts – and second the appearance of them approximately at the end of that century, according to the existing documentation, in Thera, Amorgos and Delos. A similar installation at Knossos' Palace (Angelakis et al., 2005, p. 212) is a very early example.

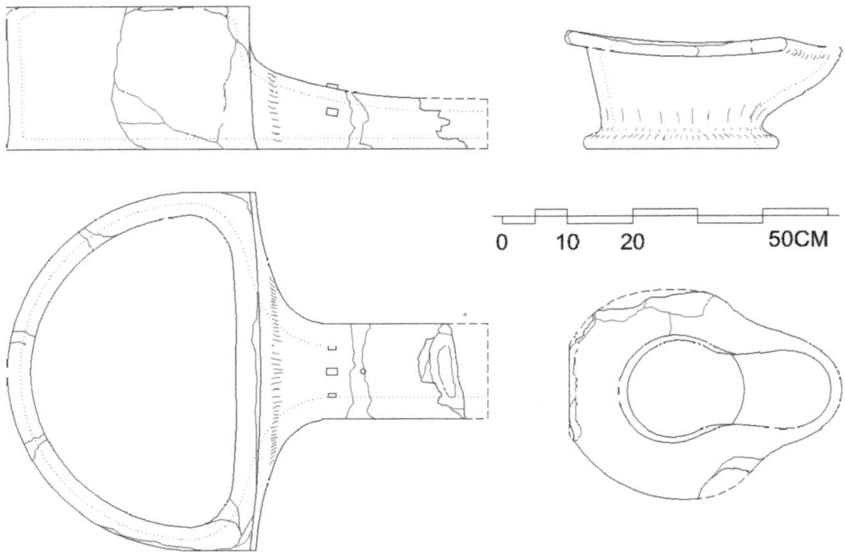

Fig. 4.1 Earthen toilet seat and defecation vessel at Olynthos (copyright permission with G. Antoniou)

Because of the number of lavatories that have been found in residences and public buildings in Delos indicates the importance of the island regarding the formation and study of ancient lavatories. Moreover the economic and social evolution of Delos during the post Peloponnesian war period as a commercial and naval center of the Helladic space justifies that importance. That society of the prosperous tradesmen and seamen was logical to confront substantially a problem that deplored all the ancient cities.

The typical layout of the lavatory was formed during the next centuries in the greater Hellenic region with numerous examples not only on the islands but also in the continental territories. Many latrines dated in the 2nd century B.C. have been preserved in residences at Amorgos (see Fig. 4.2), Delos (see Fig. 4.3), Dystos (see Fig. 4.4), Kassopi, Erythrai, Thera (Thira), and in public buildings such as gymnasiums and palestrae, such as at Asclepieion of Kos (Fig. 4.5), Miletos (Fig. 4.6), Amorgos (Fig. 4.7), Pergamon (Fig. 4.8), Phillippoi (Fig. 4.9), Epidaurus (Fig. 4.10), and Ostia (Fig. 4.11). The significance of Delos for the evolution of the typical layout of the ancient Greek lavatory is important and could be the subject of relevant detailed historical research.

The mature formation of the lavatory features in the late Hellenistic era was followed by its spread throughout the entire Roman Empire. During the 1st and 2nd centuries B.C., lavatories were built in monumental forms and sizes, equivalent to other constructions by the Romans. Figure 4.12 illustrates the lavatory outside the Roman Agora in Athens.

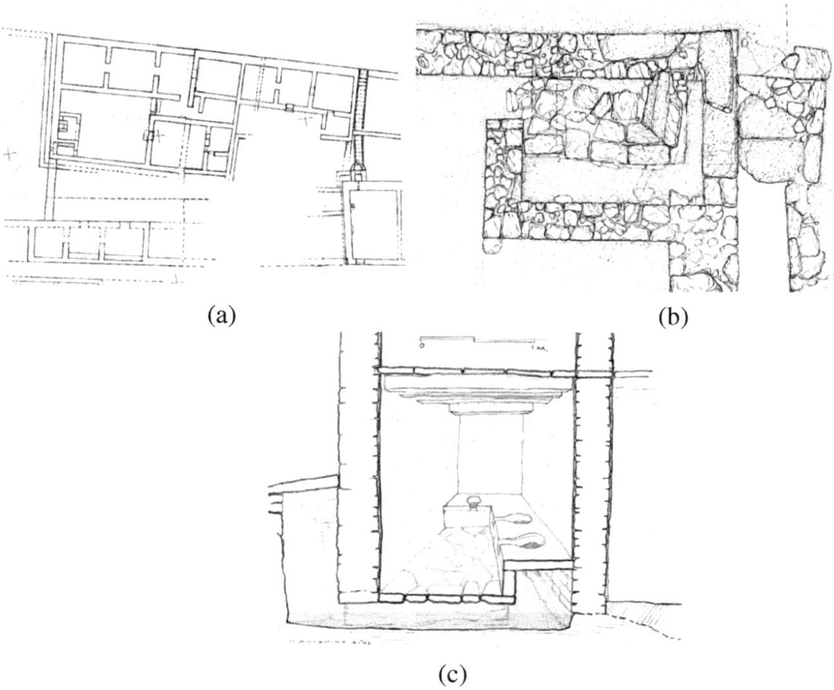

Fig. 4.2 Ithidiki's residence in Minoa Amorgas. (**a**) Plan of residence showing location of lavatory, (**b**) Plan view of Ithidiki's lavatory, and (**c**) Restored view of Ithidiki's lavatory (copyright permission with G. Antoniou)

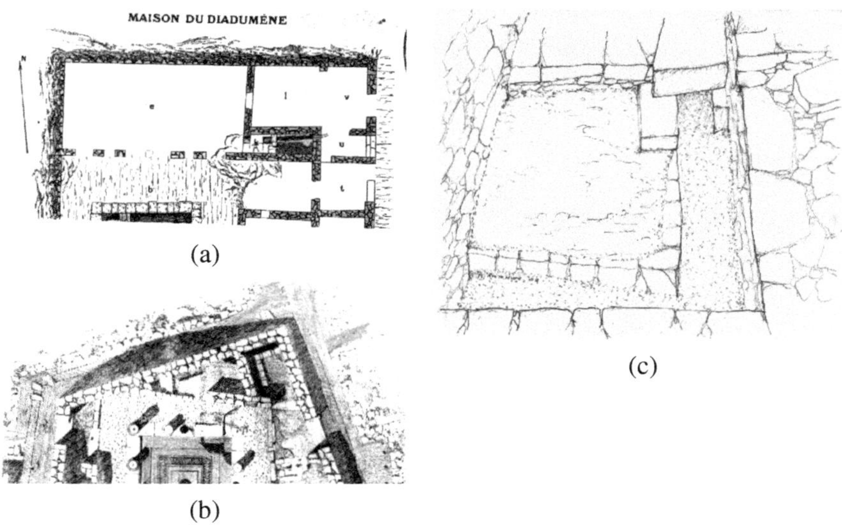

Fig. 4.3 Lavatories of houses in Delos. (**a**) with flushing hole (Chamonard, 1924), (**b**) L-shaped ditch (Chamonard, 1924), (**c**) Lavatory of house IV (copyright permission with G. Antoniou)

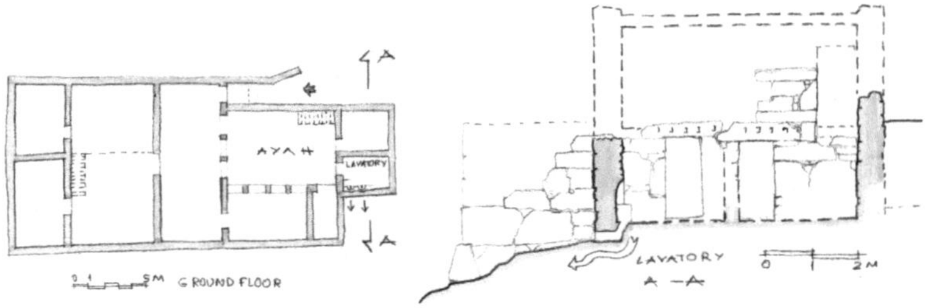

Fig. 4.4 Plan of a typical house in Dystos and the section of the ruined house (copyright permission with G. Antoniou)

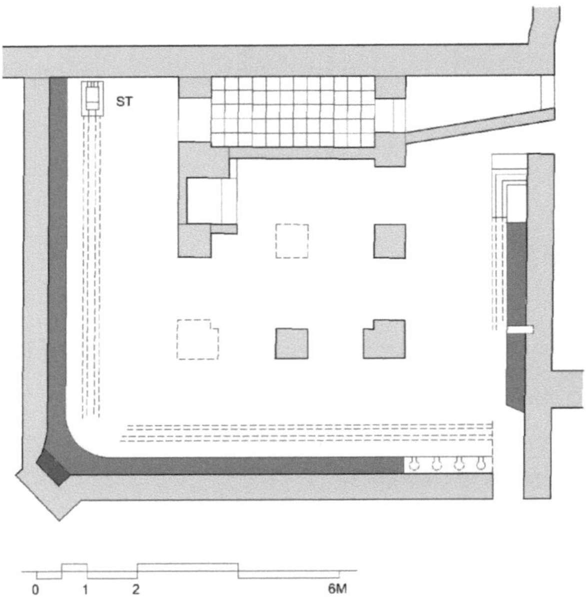

Fig. 4.5 Public lavatory in Asclepieion of Kos (copyright permission with G. Antoniou)

4.5 Description of the Typical Lavatory

4.5.1 Public and Private Lavatories

From the existing documentation it appears that the essential differences between private and public lavatories were mainly their size, represented by the number of defecation holes, and whether continuous water flow was used. In Thira there is a large number of public lavatories, whereas in Delos there is a large number of the private ones.

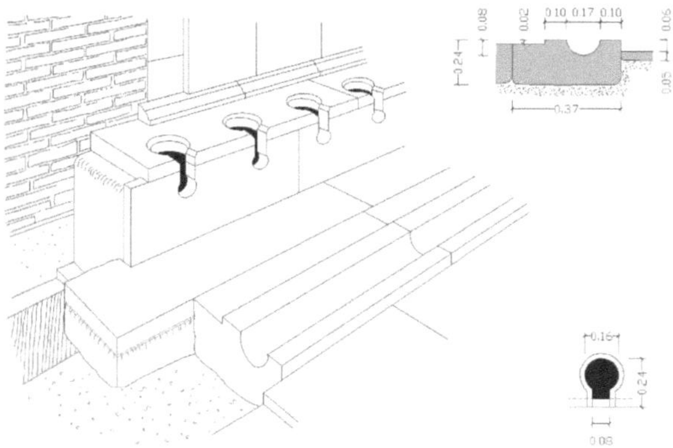

Fig. 4.6 Latrine in Miletos with dense arranged holes (copyright permission with G. Antoniou)

Fig. 4.7 Restored view of gymnasium lavatory on Amorgos (copyright permission with G. Antoniou)

4.5.2 Pipe Network – Sewerage

The layout of the lavatory was determined by the ditch under the defecation benches. Public facilities were usually supplied with water from natural flow. In many cases when natural flow supply was available for the flushing, the water was reused. In many residences the natural flow was combined with the bathroom and/or the kitchen (as in Delos and Ostia respectively). In cases where flushing was done with the reused domestic water from the kitchen or the bath, there was only an outgoing

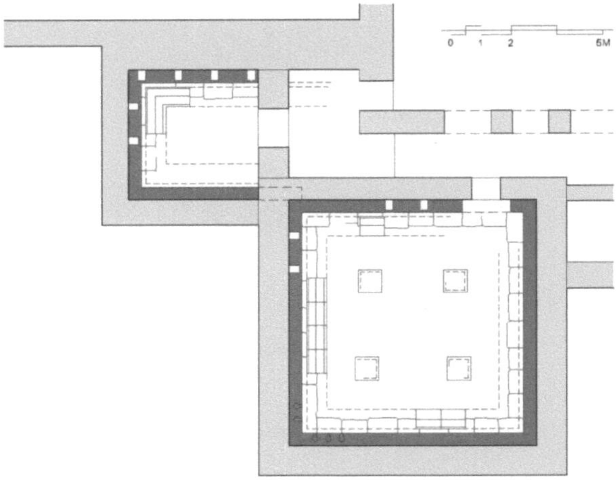

Fig. 4.8 Double public latrine at Askleipieios' sanctuary at Pergamon (copyright permission with G. Antoniou)

Fig. 4.9 Latrine at the gymnasium of Philippoi. The entrance to the latrine is between the two vertical columns (copyright permission with G. Antoniou). Color version available in Appendix

duct and a "flushing hole" at the one edge of the lavatory. Otherwise both inflow and outflow ducts were used.

Ditches and pipes defined the layout and location of the lavatory inside the building. Moreover, the requirements of sewerage placed lavatories along the perimeter of the building on the side adjacent to a street. The sewage drained through ditches along the streets or in open spaces for small houses (i.e. in Dystos). The most typical location was at the entrance of the buildings, while for residences the lavatory typically was placed in small spaces near the entrance such as in Delos. Placement

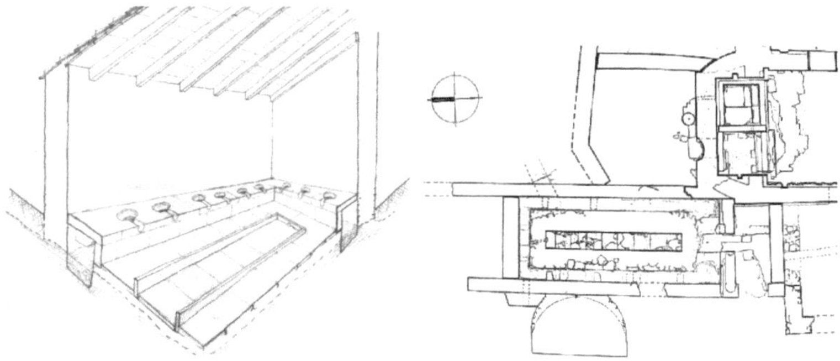

(a) (b)

Fig. 4.10 Restored views of the lavatory in Kotyo's Stoa in Epidaurus (**a**) View of seats and (**b**) Plan view (copyright permission with G. Antoniou)

Fig. 4.11 Public lavatory of Triklinon's area in Ostia (Wikipedia Commons, Al Mare)

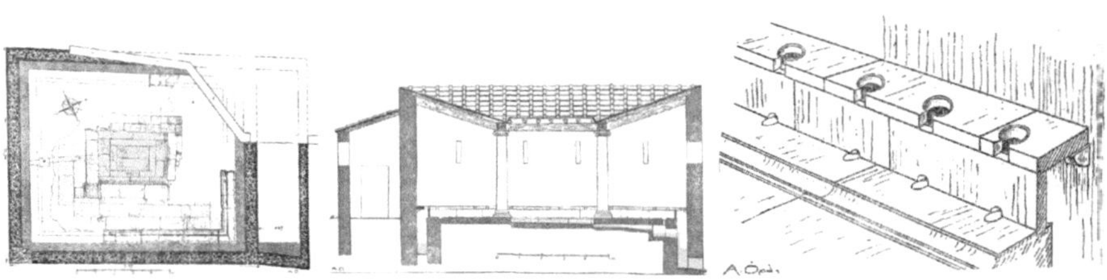

(a) (b) (c)

Fig. 4.12 Lavatory outside the Roman agora in Athens. (**a**) Plan view, (**b**) Restored view of the vertical, and (**c**) Restored view of the seats (Orlandos, 1940) of the cesspits (κόπρωνας – kopronas) were placed by the entry of Athenian houses.

The pipe network and the sewerage were reducing the chances for modifications and additions of lavatories in already existing buildings. Therefore only some buildings were equipped with a lavatory after their construction. Despite that there are some rare cases where a lavatory was added later such as at the Kotyo's Stoa in Epidaurus.

4.5.3 Typical Features of Lavatory

4.5.3.1 Perimetric Ditch

In public lavatories ditches were typically along three sides of the chamber, were arranged in a U shape, and were mostly uncovered. In smaller private lavatories the ditches were along two sides in an L shape. The later large public lavatories had ditches along the four sides and covered in front of doors (Athens, Philippoi, Asklipieion on Kos, Pergamon, and Epidaurus).

Water, which either flowed or was carried in containers, was used to facilitate the sewerage, and usually was situated opposite the sewerage duct. In many cases, like Philippoi and Epidaurus, they were constructed such that the water velocity increased and therefore improved the flushing of the lavatory. Ditches were constructed at the natural water flow level, either by adjusting the height of the lavatory as in Roman Agora-Athens (see Fig. 4.13) or by adjusting the lavatory's floor level as in Philippoi (see Fig. 4.9).

4.5.3.2 Lavatory Seats

The bench shaped seats were constructed of stone slabs, approximately 10∼20 cm thick and usually 45–50 cm wide, which was roughly the height of the seat from the floor. On the contrary the length varied depending on the size of lavatory and the number of defecation apertures on each slab and on the distance between them. These distances varied from 1.20 m at Minoa Amorgos (Figs. 4.2 and 4.13) to 2.30 m at the Philippoi's lavatory (Figs. 4.9 and 4.13). Figure 4.13 shows four types of lavatory seats and supports at various locations.

Under every lavatory seat, even the simplest ones, a vertical slab was used to cover the void between the floor and the seat. The almost standard height of 45 cm as mentioned above is as high as that of a typical chair. Seat supports have an interesting differentiation and typology. All four types of seats shown in Fig. 4.13 are cantilevered and are mostly covered except for the type at Philippoi and Efessos. More specifics are as follows:

- The cantilevered stone slab protruding out of the wall occupies two of three sides of the lavatory at the gymnasium in Minoa-Amorgos. The other 1/3 is supported by a stone bracket.

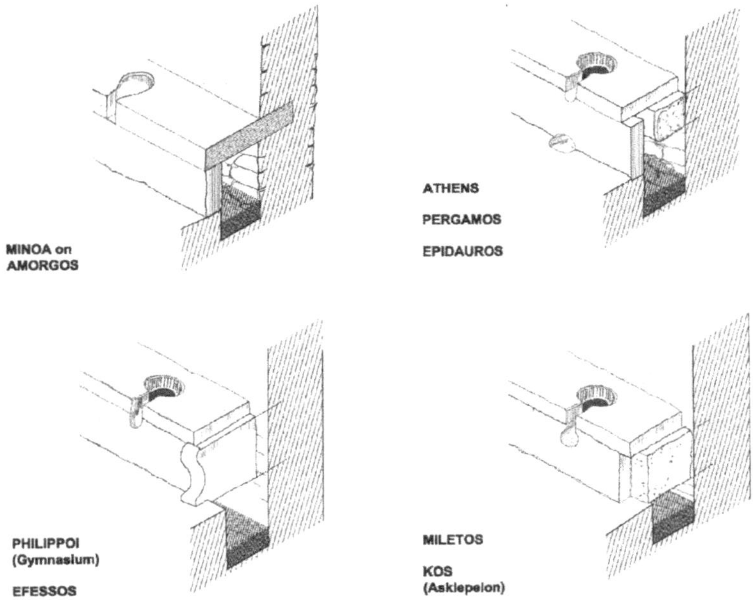

Fig. 4.13 Formation and types of lavatory seats (copyright permission with G. Antoniou)

- The freely supported slabs over the stone beams, whether cantilevered or not, are invisible. They are covered by a vertical plate which fills the void in front of the seat which is the most typical form. The joists are roughly as high as the distance of the seat from the floor. A cheaper stone than used in the rest visible structure is used. Original Roman lavatories were made of small brick pieces of wall, penetrated at their lower part by the perimetric ditch.
- Similar to the previous type the stone joists protrude out of the vertical plates and have been formed as neck mouldings of benches and exedras as at Philippoi and Efessos.
- Freely supported seat slabs were also supported by stone cantilever beams which were shorter and less wide than the seat. Characteristic examples are at the Asclepieia of Pergamon and Epidaurus.

4.5.3.3 Defecation Openings

As previously stated the distance between seats varied. In the gymnasium at Minoa on Amorgos the distance is 85 cm while at the Roman Agora-Athens the distance is just 51 cm. Obviously, at the big public lavatories which facilitated more people they were placed more densely. In most Roman latrines there was also denser arrangement of seats (i.e. in Ostia (Fig. 4.11) and Miletos (Fig. 4.6)).

Not only the shape but also the ergonomy of the openings were interesting. In fact they are more interesting than the earthen defecation seats of Olynthos. The resemblance is not only the keyhole shaped outline of the opening but also their slanted (beveled) edges. The width of the slant (bevel) in the stone openings varied from 4 cm in Minoa to only 1 cm in the Roman Agora of Athens. Possibly this slant was for an earthen cover, however one of the two known clay covers, at Philippoi, do not have a beveled edge. Probably, according to its dimensions it could be a typical short earthen drum of hot bath floor supports. Another interpretation is that this slant was curved in order to make the seat more comfortable than it would be with a sharp edge.

There was a variety in the shape of the openings. The rough but more ergonomic elliptical shape found in Amorgos, became an object of formalization in the Roman period. The prolongation of the opening up to the front edge of the seat contributes to these variations. It must be noted that the form of openings remained substantially the same during the life time of that particular type of building. Moreover it survives even today through the shape of the openings of the lavatories in the eastern Mediterranean area.

4.5.3.4 Auxiliary Elements

Other adjoining auxiliary elements were used in the expansion of lavatories. Remarkable are the small holes for drainage of urine on the floor of the lavatory in the Roman Agora in Athens but also the explicit clue on something equivalent in Minoa Amorgos, three to four centuries earlier. Moreover the small ditch with continuous water flow in a half pipe cross-section was used widely (i.e. Roman Agora in Athens, Fig. 4.12, Kos, Fig. 4.5, Miletos, Fig. 4.6, etc.). According to researchers

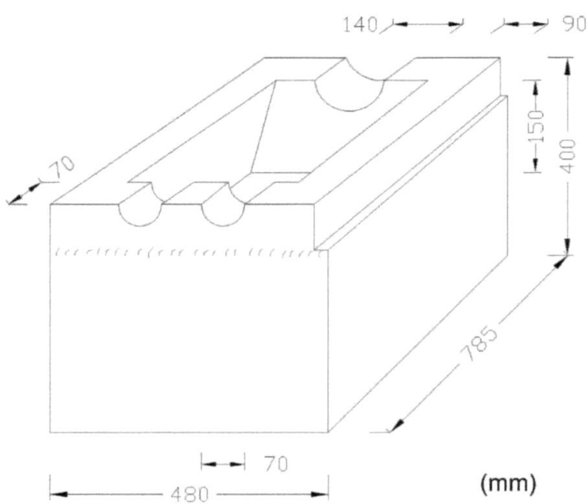

Fig. 4.14 Small sediment tanks used in lavatories at Askleipieion, Kos (copyright permission with G. Antoniou)

it was used for the cleaning of σπογγιά – sponghia, the sanitary paper of that time. In the middle of the chamber at many lavatories there was a small shallow reservoir. In Athens (Fig. 4.12) and Efessos it was surrounded by a colonnade like a Greek κατάκλειστον – *katakleiston* or a Roman impluvium. Probably it also had other uses. It had a slope of roughly 1.5% in Epidaurus. Possibly it was used for the washing of σπογγιά *(sponghia)*. A similar small central reservoir also existed in one public lavatory of Thira and perhaps had an equivalent function. Besides these elements there was a small sedimentation tank in the lavatory in Askleipieion on Kos, which also regulated the flow of the water of the small perimetric ditch. (Figs. 4.5 and 4.14)

4.5.3.5 The Layout of the Ground Plan

Most ground plans have an oblong shape in both public and private lavatories. In the known lavatories at Athens, Philippoi, Efessos and Epidaurus there was a rectangular lobby at the entrance of the narrow side and a door to the main chamber on the long side. Finally it should be noted that the evolution of the ground plan of the lavatory during the Imperial era was to ones with more complex shapes and imposing layouts.

4.5.4 Public Lavatories

One of the earlier lavatories is that of Minoa's gymnasium on Amorgos (Fig. 4.7). It is small in size with the gymnasium at its south west corner, and dated in the mid 4th century B.C. Apart from its surviving roof and the benches on three sides it is also characterized by the large conduit that supplied it with a natural flow of water. The sewerage flowed in the conduit, parallel to the south wall of the Gymnasium. It is dated probably in the end of 4th century B.C. (Marangou, 1986, 1987; Marangou, 2002; Neudecker, 1994). The lavatory is very well preserved, but only half of the monolithic floor still exists. On the other hand the door remained almost intact and only two pieces of the door jamb have fallen down.

The public lavatories of Thera are small in size but abound all over the excavated potion of the ancient city. Despite their public use, they have a small size. Figure 4.15 shows the location of a small public lavatory in ancient Thera. Even though they were residences, their access was only from the communal space of streets. Only the ditches and sewers have been preserved but not any seats or defecation openings. Possibly they were not constructed of stone. The sewerage was transported through ditches to the streets.

In Delos public lavatories have been found in the Palestras and the Gymnasium. In the Palestra of the Lake (Fig. 4.16), there are three spaces, formed after the rearrangement of the original classical building, that are attributed to that use. The neighbouring smaller and newer Palestra has a lavatory as well. In both buildings lavatories have been placed in the perimeter, and particularly near the path of some drainage. The north-eastern lavatory of the Lake's Palestra was probably supplied

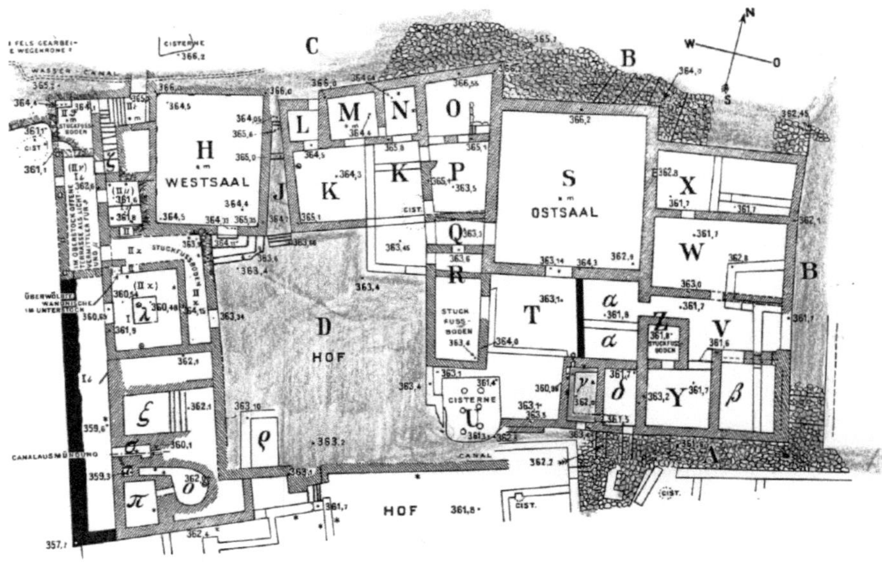

Fig. 4.15 Location of small public lavatory in ancient Thera (Gaertringen von, 1899–1909)

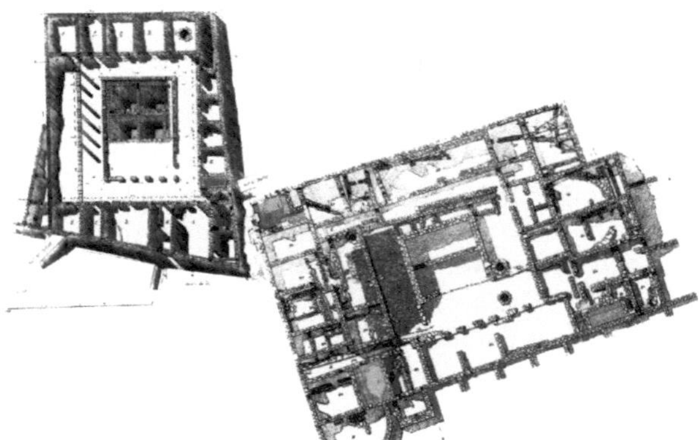

Fig. 4.16 Reconstructed planview of the palestras on Delos (Chamonard, 1924)

by the water from the bath or even the colonnaded atrium. In the south western part there was probably a small rectangular reservoir in the middle.

The lavatory of Asclipieion in Pergamon is characterized by its ground plan layout, which is more complicated than the usual rectangular form (Fig. 4.8). In the Asclipieion of Kos the lavatory is part of a later extension of the lower perimetric portico towards the west. Here a monolithic reservoir drains water from a small perimetric half pipe for the wash of σπογγιά (sponghia) to the main conduit, providing a type of water reuse within the lavatory.

At the Ventio's Thermae in Efessos the traditional Greek layout is maintained inside a typical Roman building. It is characterized by the oblong impluvium which is quite monumental for the size of the chamber. The gymnasiun of Philippoi has the typical Greek layout, despite Roman modification of the building. Also the placement resembles the lavatory of Kotyo's Portico in Epidaurus.

Two public lavatories in Athens that have been preserved date back to the Roman period. They are located in the south-eastern corner of the Attalos' Stoa and east of the Roman Agora. Both ground plans have a square shape. The Roman market's lavatory is a mature construction of that period, since it was built after the Agora. Apart from the oblong entrance lobby it is characterized by a deep conduit underneath the benches and the impluvium at the center of the room. According to the surviving parts it had 62 defecation openings, which had corresponding urinal holes on the floor. (Orlandos, 1940, pp. 251–260)

Finally in Epidaurus the lavatory at the east end of Kotyos' portico possibly could be one of the later buildings of this type in Greece. It has an oblong ground plan and is supplied with a flow of water, most probably from the north-eastern baths. Probably it was built when the portico was partly standing and the poor construction includes stones of other collapsed buildings in the sanctuary. The elongated shallow tank in the middle, made of tiles, has a small sewerage pipe that ends at the main conduit.

Three other Roman examples include the lavatory of Pompei's palestra, the complex of Triklinon's in Ostia (Fig. 4.11), and the Largo Argentina lavatory in Rome. The types of buildings were not only spread out around the Mediterranean but also throughout most of the Roman Empire.

4.5.5 Private Lavatories

The case of the remains of earthen fixed utensils (vessels) for sewerage in Olynthos could be reported as the older known residential lavatory. A common feature is placement of the lavatory at the perimeter of buildings next to the street. Most probably it was supplied by reuse of water from other household uses since no water supply conduit was found.

Numerous domestic lavatories have been preserved in Delos (Fig. 4.3). Their size is medium or small and the main ditch has an L shaped plan. The smaller ones have the bench with the openings along only one side, while the larger ones have a bench along the three sides. Most likely the seats with the keyhole shaped openings were wooden. The sewerage was applied via the conduits in the streets. Because of the lack of water on the island, they probably flushed with reused water and not with the limited water of the domestic cistern.

The resemblance of Ithidiki's residence lavatory (Fig. 4.2) in Minoa on Amorgos, with the equivalent ones on Delos is remarkable. On the other hand the main difference is that it was supplied by natural flow of water from the conduit attached to the outer wall. Most probably it carried water from the drainage of buildings at a higher elevation. The inner conduit had an L plan and the room became part of a workshop during the Roman period (possibly a glass shop). The case of the lavatory in a house of Dystos (Fig. 4.4), illustrates the effect of the terrain on the construction (Hoepfner, 1999). The inclination led to a shape and layout with similarities to the lavatory in Minoa's Gymnasium. However the main difference is that in Dystos there was no natural water flow and the sewerage flowed to the space just outside of the house, without any conduit (Fig. 4.4).

In Erythres the lavatory was placed in the corner of the atrium along the narrow side of the room, just opposite from the door. The sewage flowed outside of the building. Lavatories have been found in other ancient towns such as Kassopi behind Katagogeion. Private lavatories were spread throughout the Roman Empire similar to the public ones. In Ostia, the existence of a lavatory in almost every house was a common feature.

4.6 Operation with Reused Water

Because of the semiarid nature of many areas around the eastern Mediterranean, and the expenditures needed for cisterns and aqueducts, most ancient Greek toilets reused water for flushing. Except for the small private latrines, where the reused water was applied with buckets, in most other cases it flowed into the lavatory from natural sources, either continuously or periodically. In the last case there was a pipe network providing the reused water (see Fig. 4.17). If there was a shrine in the vicinity then that water was reused for the flushing, like the case of the lavatory at the Askleipeion on Kos. In other cases grey water was used such as the lavatories in Minoa on Amorgos, where any kind of water, residential, workshop etc., was running in the network. At the lavatory of the Kotyos Stoa in Epidaurus (Fig. 4.10b) the reused water was supplied from the therme situated 50 m north east of the lavatory.

The residential lavatories were flushed with either the left-over water from the kitchen or the bath, or the collected rainwater from the cistern of the house. In addition there is evidence that there was internal reuse of water. At the lavatory of the Askleipieion of Kos the water was supplied from the left over quantity of the shrines near the lavatory (Fig. 4.5). Then it flowed first in the perimetric ditch for cleaning the σπογγιά *(sponghia)* and then was put into the main flushing ditch under the floor.

The categories of reused water supply were as follows:

- The common type has the input of flushing water at one side and the outflow at the opposite. A subcategory of this type has the input and the output very close.
- A rare case incorporates intake of reused water and sewerage to the same duct, running just outside the building.
- A small opening in the floor was used to direct the flushing water to the under floor ditch.
- Moreover in cases of small domestic lavatories the reused water was that left over after other household uses, i.e. cooking and washing. In most examples of that

Fig. 4.17 Pipe network for reusing water in the Minoa on Amorgos (copyright permission with G. Antoniou)

type an opening on the floor, at the edge of the ditch of the toilet, was used to direct the reused water for flushing.

- Often water reuse was taking place inside the lavatory

4.7 Simultaneous Usage by Numerous People

Most of the ancient names mentioned previously refer to the lavatory as a private place (the part –από *-apo*). Despite that, all the remains of ancient Greek lavatories undoubtedly indicate simultaneous usage by more than one person. Even in residences where the inhabitants numbered at least five to ten people, there are lavatories with two to four defecation openings (i.e. in Figs. 4.2 and 4.4). In these cases it is not known if there was a simultaneous usage by residents of different sexes. At the public lavatories as well, there is no doubt that they were used by dozens of people, often more than fifty, a practice which expanded during the Roman era and survived in many Byzantine and medieval lavatories of the eastern Mediterranean, i.e. monasteries and castles as in Mytilene.

Lavatories can be classified according to the number of the defecation seats, which correspond to the number of users.

- The very small domestic lavatories used by two or three people of the house (i.e. in Figs. 4.2 and 4.4)
- Medium to large domestic lavatories with more than four defecation seats.
- Small public lavatories for at least four users (i.e. in the gymnasium of Minoa -Fig. 4.7-, and in the palestae on Delos –Fig. 4.16)
- Large public lavatories used by more than ten or twenty people, constructed mostly during the Roman period (Figs. 4.6, 4.11, and 4.12).

4.8 Conclusions

The shape, the layout, and the structural characteristics of lavatories during antiquity depended on: functional needs, anatomic requirements, and construction restrictions (because of the materials applied and the presence of water). They were similar in most of their elements over time, but also between private and public, as well as, rich and poor lavatories. The use of a lavatory by more than one individual, at the same time, remained during the life time of the toilet, which has survived to the present in the public male urinals.

Differences of ancient lavatories through the years focused in the change of their characteristics resulting in the implementation of the Roman building style, not only the size, but also the type of materials. The essential differentiation between private and public lavatories is the number of users. The appearance and evolution of such construction depended directly not only on the prosperity and the economic growth but also on technological improvement. Besides that, a similar evolution of the sanitary areas of houses and buildings has occurred in the last 50–100 years by the western type societies and it is under deployment in the third world countries.

After these remarks, the existence of numerous lavatories in the thriving Hellenistic societies around the Aegean and the wider region of the eastern Mediterranean are absolutely well expected.

In the case of Delos the large number of private lavatories is justified by the presence of affluent residents, tradesmen and seamen. However the importance in the case of Thera (Fig. 4.15), with the numerous public lavatories which, according to their placement, shape and size, could be private lavatories.

Influences from the thriving Prolemean Egypt should not be ignored. Many lavatories are reported in descriptions on papyruses related with real estate matters. Accordingly, the spreading out of the mature form of the lavatory by the Romans was expected. The morphological and institutional mutation of lavatory during that period is justified by historical facts during Vespasjan's years of rule. Lavatories became an important source of income (from entrance fees) for the imperial funds. The construction, which had been shaped during the 4th and 3rd centuries B.C., was spread around the Mediterranean by Roman engineers and generals, substantially with little changes.

The reuse of water for flushing was the result of limited water supply in most regions of ancient Greece and the eastern Mediterranean, as well as the result of expensive constructions required (aqueducts, cisterns etc.). The increase of sizes during the Roman era and the presence of natural water flow in much more public toilets should be related directly to the extensive waterworks which were constructed all around Mediterranean during that period. Today the issue of reused-or non potable water-for the toilet flushing arises in many small islands in the Aegean Sea. The increased needs for tourism as well as the expenses to build and the energy consuming desalinisation plants lead to the application of reused water for that purpose.

In addition, use of lavatories by several people simultaneously was very characteristic despite the defecation. The names in antiquity embody the sense of privacy (*withdrawal*). Single person latrines, such as at Olynthos, testify that this social habit might have appeared or evolved along with the type of the construction. On the other hand the social acceptance of nudism in the sporting games, gymnastics and later on in baths, justifies that habit. Today a similar common use exists in the male urinal rooms, but can not be considered as a continuation of that ancient habit.

References

Angelakis A, Koutsoyiannis D, Tchobanoglous G (2005) Urban wastewater and stormwater technologies in ancient Greece. Wat Res 39:210–220

Aristophanes, Hall F, Geldart W (1967) Aristophanis Comoediae. Oxonii Typog. Clarendoniano Athenische Mitteilungen, 1902, iv, 65–66

Bethe E, Pollux J (1900–37). Onomasticon. Pollucis Onomasticon e codicibus ab ipso collatis denuo. Teubner, Lipsiae.

Chamonard J (1924) Le Quartier du théâtre: Contient V. 1, V. 2 ,V. 3 Paris: 1922–1924, C. E. A. de Délos; 8, 1,2,3

Filimonos A (2000) The city waste in the rural production. The testimony of ancient Rhodes. In Lyritzis I, Sampson A (eds) Archaeology and Environment in Dodecanese. Research and cultural Tourism. International Conference, Department of Mediterranean Studies, Laboratory of Archaeometry. Proceedings, Rhodes, Greece.

Gaertringen von H (1899–1909) Thera. G. Reimer, Berlin

Hoepfner W (1999) Geshichte des Wohnes. Deutsche Verlags-Anstalt, Stuttgart, Germany

Marangou L (1986, 1987) Minoan Excavation. PAE 1986, 1987, Archaeological Society, Athens

Marangou L (2002)Amorgos 1, Minoa. Archaeological Society, Athens, Greece

Neudecker R (1994) Die Pracht der Latrine: F. Pfeil, München

Orlandos A (1940) The use of the building north of the Orologion of Andronikos Kyrristos. Proceedings of Academy of Athens, Athens, Greece.

Wikipedia, Free Encyclopedia, http://en.wikipedia.org/wiki/Main_Page

Bibliography

Antoniou G (2006) Lavatories in Ancient Greece. In: Angelakis A, Koutsoyiannis D (eds) Proceedings: 1st IWA International Symposium on Water and Wastewater Technologies in Ancient Civilizations. Iraklio, Greece, pp. 463–467

Antoniou G (2007) Lavatories in Ancient Greece. Water Science & Technology: Wat. Supply, V.7.1, IWA, pp. 155–164
Audiat J (1930) BCH 1930, 54–1, 95–130
Audiat J, Linas C, Couilloud M, Bruneau P, Ducoux H (1970) Le Gymnase. E. de Boccard, Paris
Cahill N (2002). Household and city organization at Olynthus. Yale University Press, New Haven, London
Conolly C (2002) The ancient city – The life in Athens and Rome, Athens, Patakis
Delorme J (1960) Gymnasion, De Boccard, Paris, France.
Hall F, Geldart W (1967) Aristophanes Comoediae. Oxonii Typog, Clarendoniano
Keil J (1964) Ephesos: ein Führer durch die Ruinenstätte und ihre Geschichte, Wien: Öst.Arch. Inst.
Lemerle P (1937) BCH 61, 441–478, De Boccard, Paris.
Liddell H, Scott R (1961) A Greek-English Lexicon, A New Edition rev. and augm. throughout / by Sir Henry Stuart Jones with the ass. of Roderick McKenzie. Clarendon Press, Oxford
Orlandos A, Travlos I (1986) Lexicon of ancient architectural terms. Archaeological Society, Athens, Greece
Robinson DM (1928) Excavations at Olynthus pt. 2. Architecture and sculpture. John Hopkins Press, Baltimore, USA
Robinson DM, Graham JW (1934) Excavations at Olynthus pt. 8. The Hellenic House. J. Hopkins Press, Baltimore
Schatzmann P (1932) Kos I, Asklepieion: Baubeschreibung und Baugeschichte H. Keller, Berlin, Germany
Thompson, H.A. (1960) Activities in the Athenian Agora: 1959. Hesperia 29:327–368
Travlos J (1980) Pictorial dictionary of ancient Athens. Hacker, New York
White KD (1984) Greek and Roman technology. Thames and Hudson, c1984, London, UK
Wiegand T, Schreader (1910) Priene, Berlin.

Chapter 5
Water Resource Management for Iran's Persepolis Complex

Mahdi Moradi-Jalal, Siamak Arianfar, Bryan Karney, and Andrew Colombo

5.1 Introduction

The historical patrimony of any civilization offers numerous vantage points from which to base inquiry into the rich tapestry that weaves its cultural, political, economic, technical, social and natural characteristics. An ancient culture's interactions with water, and its early attempts at managing it, provide an intriguing lens through which to observe how a society's growth and development are linked with this precious resource. Early human societies devised straightforward but elegant and innovative-technical solutions for sequestering and allocating the often limited sources of water they knew, redirecting essential quantities from rivers and other surface deposits to both urban and rural areas. In this way, the ancients established the foundation not only for their own economic and cultural development, but also for contemporary water resources engineering. Indeed, water resources management played an important role in all early urban settlements. In Iran, due to frequent drought, thoughtful attention was paid to reliable design. This is particularly evident at the Persepolis complex, the ancient capital of the Achaemenid dynasty. This monument is situated in Fars province in southwest Iran and was built approximately 2,500 years ago by the Mesopotamian civilization. Various water aspects of the Persepolis complex are presented in historical context with commentary on their present condition. Ruins of the runoff system and sewer network of the complex are analyzed to help envisage the original system and how it functioned.

At its zenith, the frontiers of the ancient Persian Empire reached their limits at the Indus River and Indian peninsula in the east, the Amu Darya River and Hindu Kush Mountains in the north-east, the Caucasus Mountains in the northwest, and the Euphrates and Tigris rivers of Mesopotamia in the west. The one essentially permanent border has always been the Persian Gulf to the south. Much of this area of more than 2.5 million square kilometers comprises the Iranian territories during the Achaemenid Empire (Fig. 5.1).

M. Moradi-Jalal (✉)
Department of Civil Engineering, University of Toronto, Toronto, ON, Canada
e-mail: mahdi@optimal-solution.ca

L.W. Mays (ed.), *Ancient Water Technologies*, DOI 10.1007/978-90-481-8632-7_5,

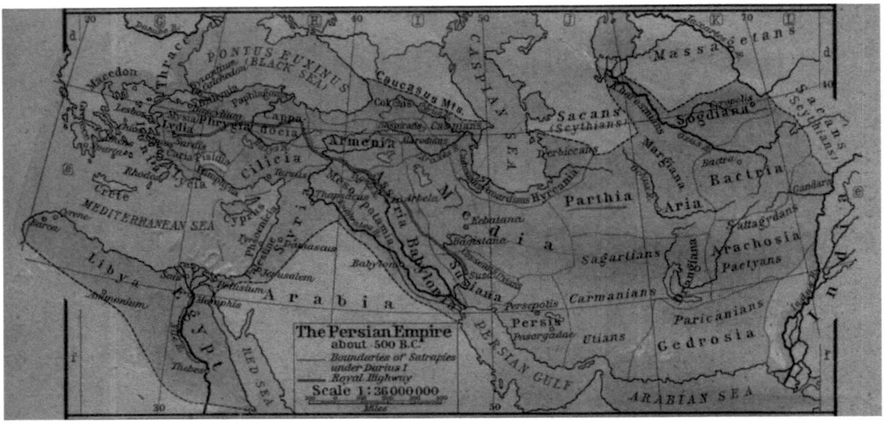

Fig. 5.1 Iran's borders in the Achaemenid era (Source: The Historical Atlas by William R. Shepherd, 1923)

The Iranian Plateau is located between 25° and 40° north latitude, largely in a zone of high barometric pressure that minimizes precipitation and increases the likelihood of droughts. The arid interior plateau is traversed by two smaller mountain ranges. Parts of this desert region are littered with loose stones and sand, with these zones gradually merging into fertile soil on hillsides. The severity of drought has been traditionally moderated for two reasons: 1 – the creation of the Caspian Sea in the north, and the Persian and Oman Gulfs in the south after water drained from the entire plateau during earlier geological eras, 2 – the emergence of alpine chains such as the Alborz Mountains which follow the contour of the south shore of the Caspian Sea, and the Zagros Mountains which run more or less parallel to the Persian Gulf. These conditions establish warm and humid areas and the combination of mountains; the northern and southern seas and the sun drenched interior prairies ensure the simultaneous experience of all four seasons country-wide. Generally, the plateau is divided into three climate zones corresponding to the hot coastal region, temperate inter-mountain valleys and plains, and the cold mountain region.

The region currently hosts about 3% of the world's population; however, because of its arid and semi-arid characteristics, with an annual precipitation of about one-third the global average, it receives a mere 1% of the global freshwater supply. By necessity, a tradition of careful water management became an integral part of all civilizations inhabiting this area.

Apart from the plateau's climate, its location along the transcontinental corridor between east and west was historically important. The routes facilitated trade, especially in luxuries such as silk, satins, musk, diamonds, and pearls between China, India and Asia Minor and the Mediterranean, extending over 8,000 km (5,000 miles). Trade along the Silk Road was a significant factor in the development of nearly all great civilizations including China and India in the east, Egypt, Persia, and Arabia in the middle-east, and finally Rome, and Byzantium in the west.

The Achaemenids (546–331 BCE) were a nomadic people comprising wandering tribes who circulated throughout the plateau. They founded various towns in which they often settled while using them as seasonal capitals. For example, they established the towns of Susa and Babel in the southwest during winter in order to enjoy the milder climate, and Ekbatana in the northwest during summer to take advantage of its cooler climate. These towns were mainly administrative, political and economic centers but, unlike them, Persepolis and Pasargad in Fars province were constructed as religious centers. While coronation ceremonies were held at Pasargad, Nowrooz, or the New Year's ceremony, took place at Persepolis and the King invited ambassadors from other countries to a festivity celebrating the Persian New Year (Ghirshman, 1961).

Persepolis is a Greek name and originates from 'Persep Tolis', a nickname of the goddess of knowledge. It was also called Parseh which means city of the Persian people and its other name is Takht-e-Jamshid (King Jamshid's throne) which was the capital of King Darius of the Achaemenid dynasty. Construction of the Persepolis complex began around 520 BCE during the reign of King Darius the great. The complex consists of several historical palaces and monuments, remains that help trace the development of civilization in the region. Following the lead of the Greeks, other Europeans referred to the city as Persepolis. It took about 120 years to construct the initial buildings and palaces which were eventually destroyed by Alexander in 331 BCE, when he conquered the Persian Empire (Shahbazi, 1976).

The Persepolis complex is located in the Marvdasht plain which was the best place to enjoy fresh and pleasant springtime weather during the new year festivities. The Marvdasht plain lies between the temperate inter-mountain valleys which are wooded and abound in streams, lakes and lush grassy meadows and was known to be excellent for grazing horses, cattle and other domestic animals. Game, including aquatic birds, was plentiful and there were vineyards and gardens of all sorts that produced a variety of fruits. The Persepolis site has the most generous groundwater resources in the Marvdasht plain, providing generally reliable water for the complex (Calmeyer, 1980).

In this regard, Sumner (1986) undertook a broad investigation of Achaemenid settlement issues in the Marvdasht plain. He argued that the sedentary population did not exceed 44,000, estimating the number of all possible villages and towns including 39 identified habitation sites and 18 possible sites in the region. His analysis also revealed a five level settlement hierarchy with the location of several named places within the Marvdasht plain.

While it is true that isolated buildings, palaces, habitation mounds and various small monuments have been catalogued in the Persepolis region, few have been the object of detailed scrutiny and only a few attempts have been made to analyze ancient systems in the region from an engineering standpoint. The aim here is to report detailed investigations of water resource management in the Persepolis region drawing from both engineering and historical sources. Regional water resource management issues include analyzing water supply systems as well as the runoff and sewer network.

5.2 Water Resource Management in Achaemenid Era

Historically, humans lived in close contact with the natural environment and satisfied their needs from their immediate surroundings, imposing a relatively small and local impact. The emergence of a sedentary civilization and concurrent improvements in agriculture led to more densely populated settlements that needed to be located on the banks of rivers or near lake shores. Archaeological evidence and extant literature clearly verify the critical need for proximity to water. For example, Ibn Khaldûn, a 14th century Arab historian, states: 'In connection with the importation of useful things and conveniences into towns, one must see to a number of matters. There is the water (problem). The place should be on a river, or near plenty of fresh water. The existence of water will be a general convenience to the inhabitants' (Khaldûn, 2005).

These 'conveniences' included water for drinking, irrigation, cleaning as well as to power waterwheels. Early water resource structures represent a milestone in the long process of human intervention in the natural surroundings.

The ancient Persian people honored the Sun (Mitra), Earth (Zam), Fire (Atra), Water (Apm Prudut) and Wind (Vahyu). These dominant natural forces were worshiped as a pantheon of deities in a polytheistic framework, a pattern echoed by the Greek and Minoan civilizations as well. Even after the emergence of the monotheistic Zoroastrian religion (600 BCE) in ancient Iran, the three basic elements of Water, Fire and Earth remained as the most important natural forces to be reckoned with and water continued to retain is venerated status in ancient Persian civilization (Ghirshman, 1961).

The ancient Mesopotamian civilizations that flourished around 6,000 years ago occupied the area of contemporary Iraq as well as the western planes of modern Iran and it is no surprise then that Iran was also party to the construction of different water resource engineering structures that are considered among the oldest in the world.

The introduction of agriculture is duly regarded as a defining moment in which humans adopted a settled lifestyle that redefined social and cultural norms, such as professional specialization, the division of labor and the emergence of hierarchies and more complex power arrangements. In order to exploit water, farmers and engineers had to deal with technical challenges, such as storage, distribution, flood control and water quality. In designing and constructing rudimentary hydraulic structures, they relied heavily on empirical observation and trial-and-error methods in order to compensate for scarce knowledge and the absence of mathematical models that their modern counterparts enjoy. When one considers that numerous aspects of hydraulic structural design have changed little over the millennia, it becomes easy to appreciate how our ancient ancestors were the trailblazers of modern hydraulics (Moradi-Jalal et al., 2006).

Ancient civilizations relied on a combination of four main water resource structures: cisterns, channels; canals, and weirs or dams. Typically, a combination of such structures would be employed in most areas. In ancient Iran, a wide variety of canals transferring water to dry areas could be found branching off from major rivers, such as the Tigris in the west and the Hirmand which courses through the southeast of the Iranian plateau. The construction of extensive and intricate water resource structures required thoughtful site selection in addition to well devised water control methods aimed at improving the land for settlement and agriculture (Moradi-Jalal et al., 2007).

Water resource management was of particular interest during the Achaemenid era and building a large network of subterranean canals was only one of several fundamental public works projects. For example, King Cyrus also built the *Ramjerd* dam on the Kur River and commissioned the *Jamshid* canals to irrigate the Marvdasht plain in Fars province. There is evidence that King Darius wrote to the governor of Godates with an order to nurture vegetation, saying: 'I appreciate your intention to improve my country by expanding fruitful trees and jungles in the north-west of Asia' (Ghirshman, 1961).

The Persepolis complex was erected on a rectangular platform of 455 m length and 300 m width, located on a foothill of the modest Rahmat Mountain. On the other side of this mountain flows a river. Because the platform's elevation is higher than the river, it was necessary to construct a small dam and build a network of canals and gutters to supply water to the complex; a main entrance reservoir was also installed in order to safeguard the precious diverted water.

The Marvdasht plain is also a high-altitude and rich flatland surrounded by the Tangeh-Balaghi Mountains. Across the plain from north to south-southwest flows the Pulvar River and it was the combination of this water source and the fertile plain that rendered the area desirable for the first Achaemenid capital. Canals were excavated to convey water from the Pulvar (Sivand) to suburb areas of the capital, and some ruins of these still exist on the northern side of Persepolis. In certain sections, it was necessary to carve masonry bedrocks for the canals, evident in the remains of the main regional irrigation canal which still exists. Considering the epoch, this 4-km canal was quite long, beginning at the Pulvar and following a steep, but declining, slope toward Persepolis where it was amenable to easier control during its approach to the complex (Nicholson and Le Strange, 1921; Feiz-Khah, 2004).

There exist remains of two earthen cascade dams 6 km northeast of Ghadr-Abad along the Pulvar River in the Didehgan valley of the Marvdasht plain which were built approximately 1 km apart. The dimensions of the first dam are 225 m long, 8 m high and with a bottom width of 35 m. It is oriented northwest to southeast in a valley where 40 m of its initial length was destroyed due to flooding and road construction. There is a reservoir behind the northeast side; on the southeast side of the dam, the remains of masonry walls belonging to a conveyance channel with cubic-shaped stones were found, stones evocative of the architectural style of Achaemenid civil works. The second dam is situated downstream of the first and is 230 m long and 50 m wide. Its current height is about 7–8 m but, due to recent damage and the absence of records confirming its original specifications, its initial height is unknown. Both dams have a clay core protected by borrow pit layers with coarse stone materials on both sides (Le Strange, 1983).

Surveys to identify material sources for palace and building construction were also undertaken and geological investigations found three significant masonry mines in the vicinity, including the Rahmat, Majd-Abad and Sivand mountain mines. Samples confirm that materials from the Rahmat and Majd-Abad mines were used most frequently for construction of the Persepolis complex. Despite the relative distance of the Majd-Abad mine, it served as the chief source of masonry materials. Although the Rahmat mine offers high strength stones, its rocks contain various faults which substantially compromise their quality. As a result, the main source of stone for Persepolis was the Majd-Abad mine (Zare, 2004).

Sumner (1986) also mentioned a site in the Marvdasht plain where there were remains of the so-called Soon irrigation system. It is not an unlikely place for rations in form of food and salary to be issued to workers engaged in canal repairs and cleaning within a 25 km radius. Presumably, the work groups camped near the canals but were supplied from other warehouses/mines rather than from the small farming villages. It is also known that tar was used for sealing canal grooves in order to safeguard against drainage beneath the Persepolis platform (Asgari, 2004), confirming that Achaemenid had almost a robust knowledge on geology, mineralogy and water resource management.

5.3 The Water Supply System of the Persepolis

The Persepolis complex is located 55 km north of the city of Shiraz, in the suburban area of the Marvdasht plain in Fars province. A stone well that was found at the foot of Rahmat Mountain revealed the nearby existence of an underground water tank that furnished potable water to the complex's palaces. Unfortunately, exploration of the water supply system of the Persepolis area, while interesting, has been at least partly neglected by archaeologists under the mistaken assumption that such inquiry was more of an engineering rather than historical interest (TCE, 2003).

Evidence supports the hypothesis that the Persepolis complex was supplied by reliable water sources including rivers, subterranean canals and springs:

I. Because precipitation was locally unpredictable, rainfall could not guarantee an adequate and reliable water supply. Furthermore, due to the challenges of storing water during the hot season, long detention times would adversely affect its aesthetic properties and taste as well as diminishing its overall quality. King Cyrus′s plan to relocate the capital from Pasargad to Persepolis, in part to build a more prestigious palace, was further motivation to devise a more reliable supply than just rain collection. Thus, in spring, water was supplied by aquifers and other underground reservoirs.
II. During the 180-year construction period of the whole complex, the large entourage of palace residents, clerks, laborers and soldiers needed water for daily living. Satisfying this need required a reliable and accessible water supply and certainly the Achaemenid experts had both the experience and skill to build this infrastructure.

Although the Pulvar (Sivand) River was the nearest permanent water source, its erratic inflow and considerable sediment load rendered it unsuitable for drinking. Moreover, in order to supply water from the Sivand River, a small diversion weir was required upstream (around *Naghsh- Rajab*area); however, no historical evidence has been found regarding the construction of such a weir. Other available local water supply resources that were exploited include the calcareous springs which were formed by limestone and Razak formations; their use is confirmed by the vestiges of old subterranean canals.

The most appropriate and accessible source is in Seidan. This village is located 19 kilometers west of Persepolis where calcareous mountains, subterranean canals and springs can be found (Fig. 5.2a), such as the Sarasiab spring with a current outflow of 400 L/s (Fig. 5.2b). Here, there is also the Hasan-Abad subterranean canal which travels along the plain towards Persepolis and Hasan-Abad village (TCE, 2003).

(a) (b)

Fig. 5.2 (**a**) High flow springs in the calcareous mountains near the village of Seidan (**b**) The Sarasiab spring in north-east of Seidan village

Nowadays, the outlet of the Hasan-Abad canal is at 1652 m.a.s.l and has a flow rate of 500–800 L/s (Fig. 5.3a). The underground canal is succeeded by an earth canal which collects all surface and subsurface flows (Fig. 5.3b) and passes from

(a) (b)

Fig. 5.3 (**a**) The outlet of Hasan-Abad subterranean canal is located in Seidan village (**b**) The earth conveyance canal of Hasan-Abad. Color version available in Appendix

Seidan and Rahmat Mountain to the west side of Persepolis, irrigating the Hasan-Abad agricultural areas, before finally reaching the city of Estakhr where it empties its surplus to the Sivand River.

Given that the source of the Hasan-Abad canal is nestled in the calcareous springs, and its production is more stable at about 500 L/s, this spring was adopted as the most suitable and reliable supply for the Persepolis complex. In the past, water flowed from this spring along the earth canal towards the northeast side of Rahmat Mountain before continuing via masonry canals to Persepolis.

Unfortunately, the earth canal was damaged by the expansion of local irrigation areas and only a small residual of the masonry canals can be seen 1,100 m north of Persepolis (Fig. 5.4) since the balance of the route was destroyed by road construction. This 80 cm wide canal has a depth of 50 cm, an altitude of 1620 m.a.s.l and follows a gradual longitudinal slope toward Persepolis.

Fig. 5.4 Remains of the masonry canal north of Persepolis

As the water was conveyed towards Persepolis, the most challenging issue was how to locally distribute it. Because of its higher elevation, the most practical method was to use underground gutters after the water entered the complex's northern platform (Fig. 5.5).

Access to required water was ensured by stairways to the gutters, while surplus water was transferred through underground canals to the southeast corner of the platform from where it was distributed to downstream irrigation areas. The entrance canal had 6 meters depth and its bottom elevation was 1625 m.a.s.l. The bottom elevation of northern platform is 1618.5 m.a.s.l (Fig. 5.6) which is lower than the bottom elevation of the entrance canal. This bi-level section serves as a small detention structure to partially regulate incoming flows before distribution. The schematic of the Persepolis water supply system is shown in Fig. 5.7.

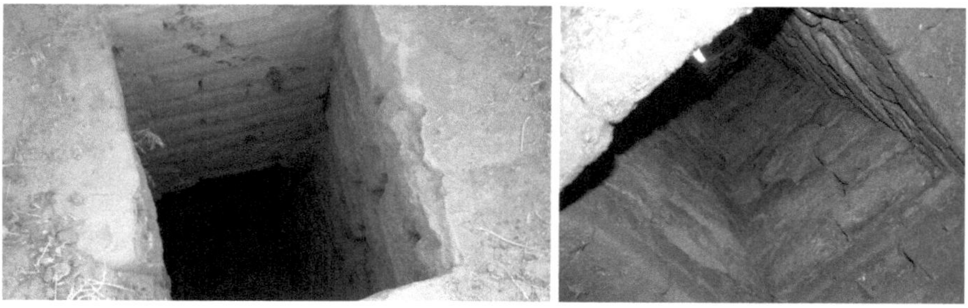

Fig. 5.5 Underground gutters at the northern platform

Fig. 5.6 Location of northern platform and entrance canal

5.4 The Runoff and Sewer Networks of Persepolis

The Persepolis complex was constructed at the foot of Rahmat Mountain, a site marked by a 125,000 m^2 terrace which is partly artificial and partly hewn from the mountain (Fig. 5.8). As the Persepolis complex was the religious capital of the Achaemenid dynasty, most ceremonies were held during winter's end and early spring when the region enjoyed greater precipitation. It was at this time that the complex's runoff network assumed importance.

In order to prevent flooding, runoff from Rahmat Mountain over the site area was controlled according to two major approaches: (1) Mountain runoff was conveyed toward a 60 m deep reservoir (a square-section well 4.2 m on each edge) located south of the site (Fig. 5.9a) with the runoff conveyed along masonry gutters to the reservoir (Fig. 5.9b) and (2) A 180 m long conduit, with 7 m width and 2.6 m depth, located just west of the site, also directed excess water away from complex buildings. If the runoff exceeded the reservoir's capacity, it was spilled downstream to the western plains (TCE, 2003).

Because it had to protect the royal buildings, the sewer system within the complex was more intricate than the excess runoff diversion scheme just described. The

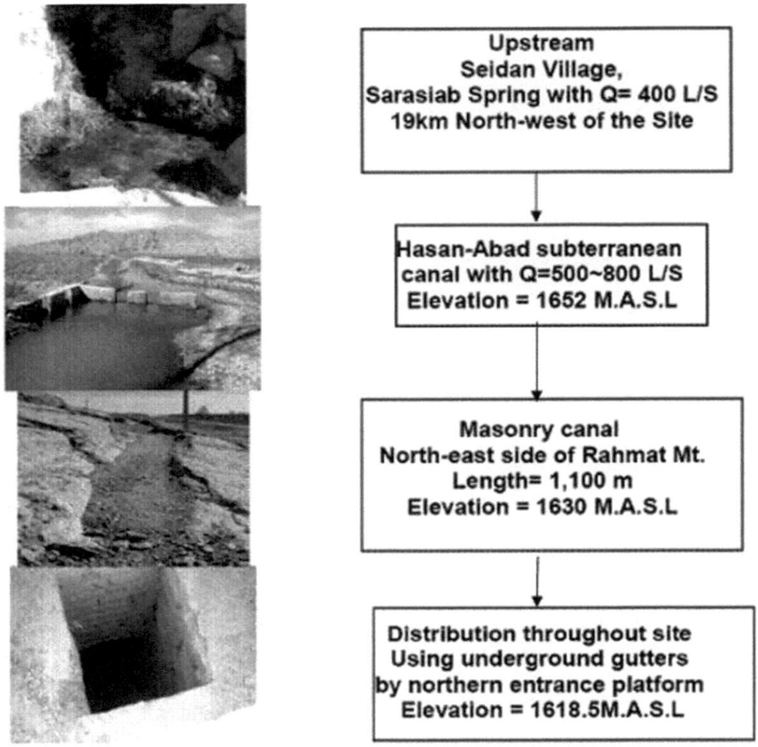

Fig. 5.7 Schematic of the Persepolis water supply system

Fig. 5.8 Aerial view from Rahmat mountain of a virtually reconstructed Persepolis

entire complex, which was served by closed section conduits and gutters, is divided into northern and southern sides, comprising five zones in total (two for the northern side and three for the south). The zones were based on the actual locations of particular palaces and buildings (Fig. 5.10).

(a) (b)

Fig. 5.9 (**a**) Temporary reservoir (well) for runoff, (**b**) Masonry runoff gutters

Fig. 5.10 North and South sides with main zoning

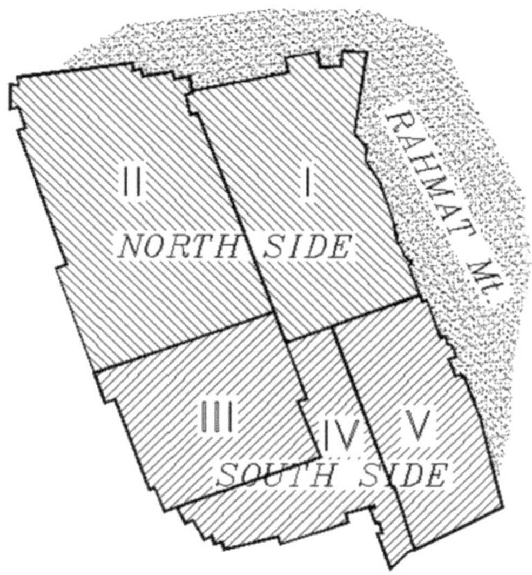

Galleries convey runoff to secondary underground gutters which are then connected to the main sewer conduit. It should be noted that the layout of the sewer networks within all five zones was designed to maintain the greatest harmony with palace architecture and the complex's scheme. For example, runoff from palace roofs, was initially transferred vertically through small holes at the center of columns down to the lower floors and then conveyed horizontally in embedded floor grooves before being directed to the main galleries beneath the ground floor. In order to facilitate runoff circulation within the network, the secondary elements were connected to the main gutters in such a way that the bottom elevations of the secondary (small) galleries were higher than the bottom elevations of the main (big) galleries (TCE, 2003).

5.5 Rainfall-Runoff Evaluation of Persepolis

Estimation of regional precipitation was undertaken using data from the nearest climatologic station. Table 5.1 summarizes average monthly, seasonal and annual precipitation data from the Persepolis station. Use of modern data obviously assumes that current climate is not very different from the climate of that period.

Table 5.1 Precipitation data from the Persepolis climatologic station

month Parameter	Jan.	Feb.	Mar.	Apr.	May	Jun.	July.	Aug.	Sep.	Oct.	Nov.	Dec.	Annual
Avg.(mm)	70	60	60.7	30.1	6.1	0.2	0.09	0.3	0.0	5.9	19.5	68.9	323.9
%	21.7	18.5	8.7	9.3	1.9	0.1	0.6	0.1	0.0	1.8	6.0	21.3	100
Season		Winter			Spring			Summer			Autumn		
Avg.(mm)		141.0			36.4			2.2			94.2		323.9
%		59.0			11.2			0.7			29.1		100

For the next step, the Gumble distribution was selected based on the smallest standard deviation needed to estimate the maximum 24-hour intensity for different return periods (listed in Table 5.2). Figure 5.11 presents

Table 5.2 Gumble estimates of different return periods for 24-hour rainfall intensity

Return period (year)	2	5	10	20	25	50	75	100	200	1,000	2,000	10,000
Precipitation (mm)	38.6	56.9	69	80.7	84.4	95.8	102.4	107.0	118.3	144.3	155.6	181.6

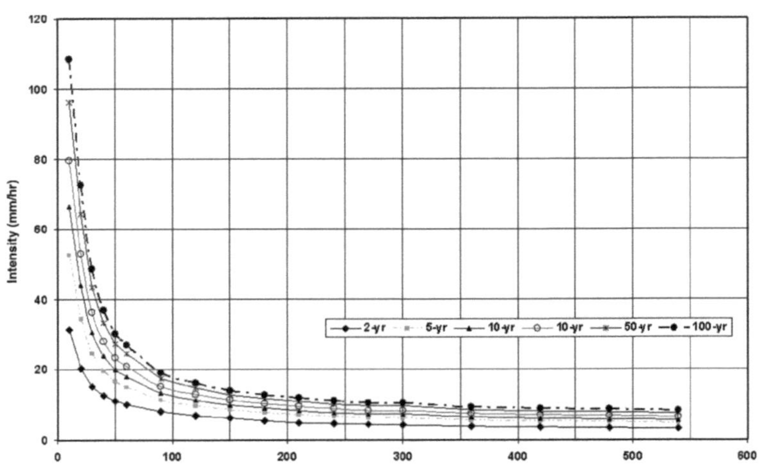

Fig. 5.11 IDF curves of different return periods for the regional station

Table 5.3 Runoff volumes of 24-hr rainfall with required/existing section areas

Zone		Area	Intensity	Runoff Coef.	Max Disch.	Required section	Existing section
		m^2	mm/hr	C	Lit/Sec	width × depth	cm × cm
Outside area		138,775		0.70	410	80×80	700×260 + 2×60×80
Main site	Zone I	42,565	15	0.50	90	60×25	60×120
	Zone II	30,115	15	0.50	60	60×20	60×100
	Zone III	17,394	15	0.50	40	60×14	80×100
	Zone IV	15,053	15	0.50	30	60×13	80×120
	Zone V	19,768	15	0.50	40	60×13	60×120
Total		263,670	–	–	670	(Slope=0.1%)	

intensity-duration-frequency (IDF) curves corresponding to different return periods which are extracted based on the regional synoptic station in the site area (TCE 2003).

There are various practical approaches for estimating runoff including rational, Kook and curve-number methods. For this catchment (0.26 km^2), the rational method was selected. Table 5.3 lists runoff estimates for the external area and all five internal zones for a 50-year return period.

The two rightmost columns of Table 5.3 allow for comparison between the existing galleries with required cross-sections assuming 0.1% longitudinal slopes and applying open channel flow formulas. Clearly, the galleries were large enough to easily convey runoff associated with a 50-year precipitation event. It is easy to imagine that these galleries were used not just for runoff, but also for the water supply system, residential sewer network and for conveying irrigation water to the downstream areas.

Excavations have revealed that sediments partially filled all major gutters and galleries, thus a rehabilitation plan was implemented to extract sediments from the sewer network and to ascertain the functionality of the old runoff system. Since the runoff volume for all inside zones (I–V) is estimated at 260 L/s, it is clear the capacity of the old systems were much further than what was required just for runoff collection (Fig. 5.12).

5.6 A Brief Reflection

The great challenge of historical work in general, and with this project in particular, is so much must be guessed, interpolated or surmised. Too often one does not have the benefit of a well-developed cultural context to fit the pieces into a great archeological, pattern of the way of life that connects the structures to the people that used them.

Perhaps even in this struggle, a water analogy is again suitable. Michael Ondaatje, in his intriguing 1989 poem 'Walking to Bellock', pictures two figures

Fig. 5.12 Existing runoff galleries in different zones

wading in an opaque stream over a complex bottom. They struggle along, sometimes bumping into guessed-at obstacles. The poem is a fitting image to historical work, and two stanzas are worth quoting in full:

> The two figures are walking
> as if half sunk in a grey road
> their feet tentative, stumbling on stone bottom.
> Landscapes underwater. What do the feet miss?
> Turtle, watersnake, clam. What do the feet ignore
> and the brain not look at, as two figures slide
> past George Grant's green immaculate fields
> past the splashed blood of cardinal flower on the bank.
> Rivers are a place of philosophy but all thought
> is about the mechanics of this river is about
> stones that twist your ankles
> the hidden rocks you walk your knee into –
> feet in slow motion and brain and balanced arms
> imagining the blind path of foot, underwater sun
> suddenly catching the almond coloured legs
> the torn old Adidas tennis shoes we wear
> to walk the river into Bellrock.

And thus is the Persepolis complex known – invitingly, intriguingly – and yet much remains unknown, its detail only guessed at, obscured both by time and by cultural distance, known to us more by feel than by certainty.

5.7 Conclusions

The ancient Persepolis complex is considered a critically important historical complex in Iran, indicative of a developed civilization. It is therefore important to explore its history as well as to protect it. A significant recommendation would be to restore its essential historical function, and to learn more of its ancient use and cultural setting.

Rehabilitation and repair of the water supply network and the sewer network are required, the underground gutters need to be cleaned and sediments and mud removed, especially at the incomplete entrance gate area and at the exit in the south- east side of the Persepolis platform. If water were to flow from the Sivand River through the Seidan plain, the system could be once again brought to life, and Persepolis would itself be further enriched as the inspiring and significant heritage site it already is.

Acknowledgments The authors thank Enr. Ali Mobini, president of TCE for his kind cooperation.

References

Asgari A (2004) Achaemenids used sealing for water canals. Cultural Heritage News Network, http://www.chn.ir/News/?section=2&id=15721 , Cited 19 Dec 2004

Calmeyer P (1980) Textual sources for the interpretation of Achaemian palace decoration. Iran 18:53–63

Feiz-Khah M (2004) Exploring water supply canal in Persepolis. Cultural Heritage News Network, http://www.chn.ir/News/?section=2&id=27775 , Cited 24 May 2004

Ghirshman R (1961) Iran from the earliest times to the islamic conquest. Penguin Books, Original from the University of California, 368 pp

Ibn-Khaldûn I (2005) The Maqaddimah, an introduction to history. Transl. F. Rosenthal, Princeton, NJ: Princeton University Press, 465 pp

Le Strange G (1983) Description of the province of Fars in Persia at the beginning of the 14th century, from Nuzhat-al-Kulub of Hamd-Allah Mustawfi, Published by Dunyā-yi Kitāb, 110 pp. (Originally published: Leiden: E.J. Brill, 1915)

Moradi-Jalal M, Colombo AF, Karney BW (2006) Conservation and rehabilitation of Iran's ancient Bahman weir, 1st IWA Int. Sym. on Water & Wastewater Technologies in Ancient Civilizations, Iraklio, Greece, 28–30 Oct 2006

Moradi-Jalal M, Karterakis SM, Karney BW (2007) Water supply for Iran's Persepolis complex, 5th International Symposium on Environmental Hydraulics (ISEH V), Tempe, Arizona, USA, 4–7 Dec 2007

Nicholson RA, Le Strange G (1921) The Farsnama of Ibnu 'L-Balkhi. Luzac and Company Publication. London, 199 pp

Shahbazi AS (1976) Persepolis illustrated, Institute of Achaemenid Research Publication, No. IV, second edition, Tehran, Iran

Shephard WR (1923) The historical atlas. Henry Holt & Company, New York

Sumner WM (1986) Achaemenid settlement in the Persepolis plain. Am J Archaeol 90(1):3–31
TCE (2003) Investigation of water supply and surface drainage network, Tamavan Consulting Engineers Co., Technical Report, Tehran, Iran
Zare A (2004) Builders of Persepolis were expert geologists, Cultural Heritage News Network, http://www.chn.ir/News/?section=2&id=27775 , Cited 21 Dec 2004

Chapter 6
A Web Based Information System for the Inspection of the Hydraulic Works in Ancient Greece

Nikos Mamassis and Demetris Koutsoyiannis

6.1 Introduction

Most ancient civilizations exploited water resources constructing hydraulic works, in order to support the everyday water needs. Ancient societies that flourished on the Greek territory since 3000 B.C. except their great contribution to philosophy, politics, sciences and arts, constructed several technical works. Many of these structures were related with water use. Using hydraulic technologies combined with understanding of processes, ancient Greeks supported several needs such as water supply, drainage of the lands and the cities, flood protection, sanitary facilities and even water use for recreational purposes. Their familiarity with water use is depicted in the hydria of the 6th century B.C. (Fig. 6.1), where young men take baths in a public installation.

The Ancient Greek societies held an exceptional position to the use and management of water recourses as testified by: (a) the advanced technologies they applied in the construction of several hydraulic works; (b) the sustainable water management practices they adopted; (c) the high living standards related to water use they developed; and (d) the explanations about the natural hydrometeorological phenomena they devised.

As a result of thousands of years of creative activity, several simple hydraulic structures (cisterns, lavatories, wells, aqueducts) or more advanced ones (dams, tunnels, siphons) are spread all over the wider ancient Greek territory. These works supported several water uses as listed above. Their presence reveals that ancient Greeks wisely resolved several problems concerning water that modern societies still have to face. Also in the ancient Greek literature there is a plethora of references about: (a) sustainable water management practices, (b) hydraulic works that are not preserved to date and (c) impressive exegeses about hydrometeorological

---

N. Mamassis (✉)
Department of Water Resources School of Civil Engineering, National Technical University of Athens, Heroon Polytechneiou 5, GR 15780, Zographou, Greece
e-mail: nikos@itia.utua.gr

L.W. Mays (ed.), *Ancient Water Technologies*, DOI 10.1007/978-90-481-8632-7_6,

Fig. 6.1 Young men are bathing (6th century B.C.; drawing by V. Kordoni based on a theme of a black-figured hydria kept in Leiden, National Museum of Antiquities)

processes (e.g. evaporation, condensation, hail, snow, rain). During the last decade, hydraulic works and water management practices developed in ancient Greece have been revisited with an increased interest. Several researchers originating from different scientific fields (archaeologists, engineers) studied the (relatively unknown) water technologies, hydraulic works and water practices developed in ancient Greece (Angelakis et al., 2005; Koutsoyiannis et al., 2008; Mays et al., 2007). The Greek philosophers' explanations about the related natural phenomena have also been studied (Koutsoyiannis et al., 2007). At this time a considerable amount of this knowledge can be found in papers published in scientific journals and conference proceedings.

To facilitate the inspection of available information about the hydraulics works in Ancient Greece a web based application has been developed. The application includes the necessary informatics tools to manipulate and analyze the various information types and make the information available on the Internet. Information includes geographical location, technical characteristics of the structures, drawings, maps, texts, papers, studies, photos, videos etc.

The main purposes of this application are: (a) gathering and archiving of all available information which is characterized by lack of homogeneity, (b) codification of the above information, (c) facilitation of its analysis using informatics tools to perform queries or to make maps and (d) an easy access from the general public and researchers to all available information.

The aim of this system is not to develop sophisticated informatics tools, but mainly to create a basic information pool concerning ancient Greek water knowledge. In order to serve this task continuously, the system must be enriched and be extended gradually, incorporating new findings.

6.2 Information System

The web based information system (http://www.itia.ntua.gr/ahw/works/) includes the necessary informatics tools to manipulate and analyze the various information types and also make the processed information available on the Internet. Open source technologies were used for the development of the various applications. The operating system is Debian GNU/Linux running in an Apache web server, the database is PostgreSQL, the programming language is Python and the web application framework is Django.

The general scheme of the information system is presented in Fig. 6.2. The system consists of a Database (DB), a Geographical Information System (GIS), a Digital Library (DL) and a Website that integrates the entire system. Also there are two separate levels of access through the web. At this time the application under development, although it already contains rich information as detailed below.

The GIS includes the geographical location of each structure and is related to the DB to perform queries and make maps. Using this, the researcher can have a wider perception of the ancient hydraulic works. The distribution of the structures in space can be combined with other characteristics such as the construction period, the type of the structure and the specific sociological and political characteristics of the society in the period of construction. In Fig. 6.3a the geographical distribution of all hydraulic works with respect to the construction period, is presented. Obviously the areal distribution of the hydraulic works is strongly related to the region where each civilization flourished. The structures are concentrated in three specific areas: Crete Island, Peloponnesus and Athens, the cradles of Minoan, Mycenaean and classical Greek civilizations, respectively. In Fig. 6.3b the sites of aqueducts have been extracted and are presented. The locations of the aqueducts are relatively dispersed from the birth places of Greek civilizations, but later Romans built them in every place of their empire, which included the Greek territory.

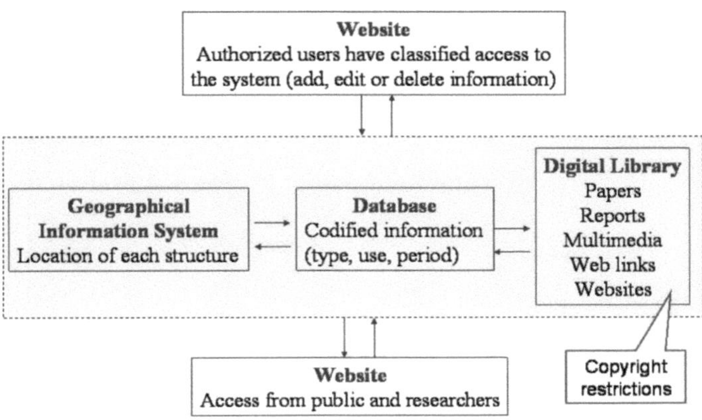

Fig. 6.2 General scheme of the application

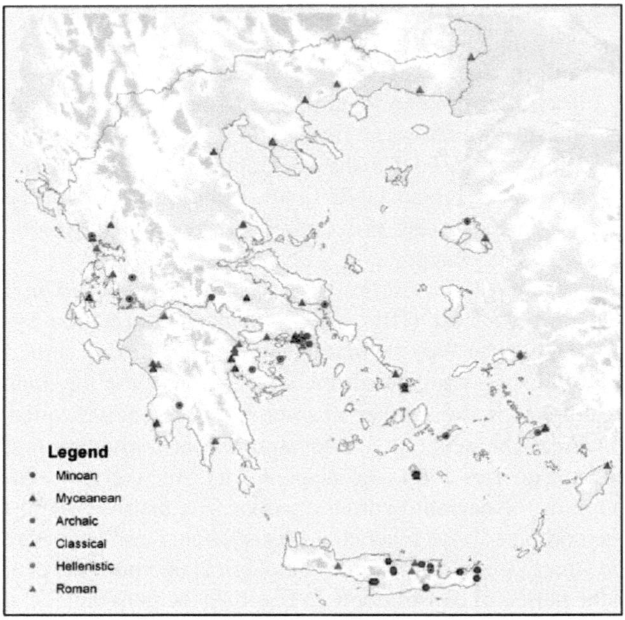

(a)

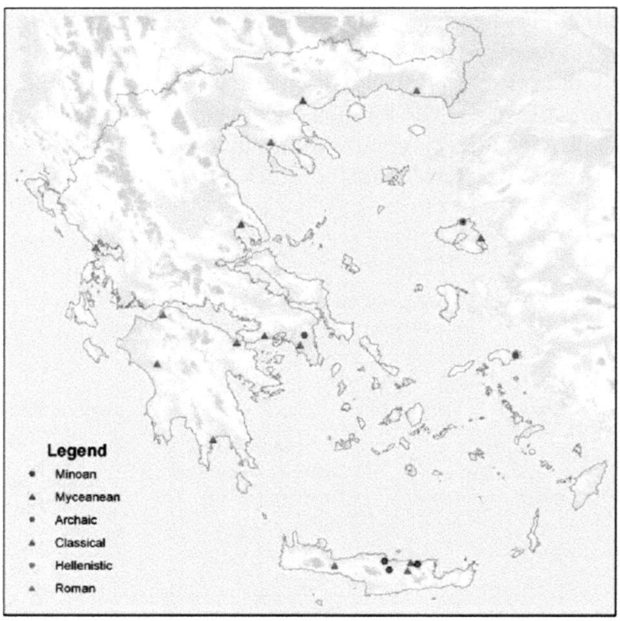

(b)

Fig. 6.3 Examples of maps constructed by the system: (**a**) Hydraulic works classified by construction period, (**b**) aqueducts classified by construction period

In the DB the organized information about each hydraulic work is stored. The main table of the DB contains fields such as: (1) name, (2) region, (3) type (main – secondary), (4) use (main – secondary), (5) period of construction, (6) name of hydrosystem, (7) brief description, (8) today's condition of the structure, (9) remarks and (10) related documents.

The region includes the name of the site and the Greek area. The types of the hydraulic works (main or secondary) can be: aqueducts, dams, tunnels, cisterns, lavatories, canals, siphons, river control works, fountains, sewers, agricultural drainage works and urban drainage works. The use of the hydraulic work (main or secondary) can be: urban or agricultural water supply, urban or land drainage, flood prevention and urban sewerage. The construction period of each work can be: Minoan and Cycladic (3500–1200 B.C.), Mycenaean (1600–1100 B.C.), Archaic (about 800–500 B.C.), Classical (500–336 B.C.), Hellenistic (323–146 B.C.) and Roman (146 B.C.–323 A.D.). An extension to the Byzantine period (323–1453) has been planned. Also another 'virtual period' (Mythology) has been created to include several hydraulics works described in myths. These myths come from the prehistoric period and exist in ancient Greek literature. Most of them refer to the labours of Heracles (Hercules) and describe several river regulation works, including river diversion, and land reclamation. There is a specific field for the name of the hydrosystem that the structure belongs. This field is very essential for the inspection of different structures that worked together. Also there are fields with a brief presentation of the structure and its current condition. Finally the most essential field is the connection with the digital library where all related documents with the structure are gathered.

Today the information system contains information about 100 hydraulic works and 20 important hydrosystems (still there are other sites under study) but the database is being enriched continuously. The distribution of the registered hydraulic works and hydrosystems through different time periods is presented in Table 6.1.

Table 6.1 Time distribution of hydraulic works and hydrosystems registered

Period	Hydraulic works	Important hydrosystems
Minoan	28	Knossos, Phaistos, Mallia, Zakros, Akrotiri
Mycenaean	11	Mycenae, Tyrins, Pylos, Copais, Olympia
Archaic	7	Athens, Samos, Archaic Thera
Classical	21	Amphipolis, Megara
Hellenistic	16	Pergamos, Messene
Roman	14	Corinth, Patras, Mytilene, Nicopolis, Olynthus, Palaiopolis
Total	97	

Currently the database contains important hydraulic works, but lacks information related to smaller and more usual structures like wells, cisterns and fountains. On the other hand almost all important hydrosystems of the Greek ancient world are represented in the database by a number of structures. Through the centuries many of these hydrosystems displayed a continuous transformation or upgrading as the local society changed its priorities, perceptions and values. The hydrosystem of Athens is such an example because of its continuous operation until the 20th century.

In the DL the large amount of information (in several formats) that is related with each structure, is stored and managed. This information includes: (1) scientific papers (or their web links), (2) reports with technical characteristics, (3) drawings and maps, (4) multimedia (photos, videos, movies), (5) related web sites, (6) references of the structure from classical texts and (7) references of the structure from scientific papers. Using a web browser the researcher can access the other three subsystems (Database, GIS, DL). Part of the information in DL is under the Copyright Law, so at this time this part is not visible through web pages to the general public.

Currently the application does not include features such as the following, which however are scheduled for incorporation in a future expansion of the system:

- Harbours, for which exists the application Limenoscope, (Theodulu and Memos, 2007). This is a database operated through the web (http://www.limenoscope.ntua.gr/) with main aim the dissemination of harbour related information.
- Water management practices that were applied in several Greek cities. Their descriptions can be found in the classical literature, and in many cases in text related to the legislation.
- Explanations of natural phenomena and physical theories that had a relationship with the hydraulic works, because they triggered the advancement of technology. These include the foundation of hydraulics, by Archimedes and Hero (Heron) of Alexandria.
- Hydraulic mechanisms that were invented in the Greek antiquity. An exceptional example is the pump (screw) of Archimedes, a device that is still in use up today.

6.3 Historical Evolution of Hydraulic Works

The information collected so far allows us to summarize the history of the Greek hydraulic technologies, which started from Bronze Age and after over 2,000 years of development, were inherited to the Romans. The evolution of hydraulic works in Ancient Greece is strongly connected with the different characteristics of Greek civilizations. The small scale hydrosystems of the Minoan villas and Mycenaean fortified palaces were gradually ruined and replaced by larger systems during the Archaic and Classical periods. The structures of Hellenistic and Roman period incorporated new advanced water technologies and construction techniques, resulting in hydraulic works that in many cases are still in use. Romans who conquered Greece at the end of the second century B.C., respected the local culture and science and improved the current technology in order to support their vast empire with a variety of technical works.

Reviewing the material of the database we can summarize some special characteristics of each period, related with the scale, the type, the use and the visibility of the structures

The majority of the recorded structures of the Minoan and Cycladic periods are located in the four palaces of Crete (Knossos, Phaistos, Mallia, Zakros). In these well organized sites several small scale hydraulic works (aqueducts, cisterns, wells) operated to ensure the water supply and drainage of their dwellers. Also small scale facilities (lavatories, bathtubs, recreational fountains) ensured a luxury way of life that can be compared with the modern one. During this period small scale hydraulic works can be also found in other places of Crete (Archanes, Chamaizi, Hagia Triadha, Palecastro, Pyrgos, Tylissos), in the islands of Thera and Cyprus, and in Asia Minor.

The water facilities of the peaceful Minoan civilization were also followed by the Mycenaean warriors. The fortified palaces of the Mycenaean period (Mycenae, Tyrins, Pylos) in mainland Greece are also places that hydraulic technologies were applied. During that period the larger scale water management for agricultural purposes (irrigation, drainage) led to the construction of hydraulic structures that control larger water quantities. A great example of such a work is the drainage of the Lake Copais located in Boeotia. The water of the Boeotic Kephisos River that fed the lake, were directed through a 25 km canal to natural sinkholes in the karst subsurface. The constructors (Mynians) protected the reclaimed land with dykes and used it for cultivation but also for building a palace in a former small island of the lake. The project was in operation until the collapse of the Mycenaean civilization. Other drainage systems of this period are those of Tiryns and Olympia with the first system to include a 10 m high dam in its structures. The dam was used to divert waters from one stream to another, perhaps to protect the city of Tiryns from floods.

During the Archaic period hydraulic works similar with those of previous periods were constructed. The higher populations of the cities increased the need of transferring water from distant sources, in addition to local water resources (wells, cisterns, springs). During this period the engineering experience of the past matured, enabling the construction of more advanced technical works. Two very important aqueducts, the Peisistratean and the Eupalinean, are the main contribution of the Archaic period to the water technology. The Peisistratean aqueduct that transported water to the city of Athens was a huge technical achievement. The main part of its 7.5 km route was constructed as a 14 m deep tunnel (for security reasons), in which a ceramic pipeline was laid.The Eupalinean aqueduct, that transported water in the city of Samos, comprised the fist deep tunnel in history dug from two openings. The tunnel was constructed around 530 B.C. with a length of about 1,050 m. Eupalinos had a great knowledge of engineering, which enabled the digging of the tunnel from two openings, a practice followed also today. He used advanced geometrical techniques to eliminate the impact of uncertainty in position in the tunnel excavation and ensure the hydraulic gradient to sustain flow in the aqueduct (Koutsoyiannis et al., 2008).

During the Classical period a vast variety of hydraulics works were constructed to serve the urban water supply and sewerage of the prosperous Greek cities. In many cases smaller structures (cisterns, fountains, wells) supplemented larger hydraulic works (aqueducts), augmenting the water management possibilities and forming the concept of a large scale hydrosystem. During this period Athens, the most important city of the Greek antiquity, had a population of more than 200,000 and an extended hydrosystem. This was based mainly on the Peisistratean aqueduct, two rivers with ephemeral flow, natural springs, and wells and cisterns supplied by storm water.

Other important hydraulic works of this period were a dam in Alysia (western Greece) and an aqueduct that carried water in a water tank in the city of Megara, near Athens. The dam of Alysia is preserved in good condition (Fig. 6.4) and is characterized by its masonry body and the spillway that operates still today.

Fig. 6.4 Alysia dam: (**a**) General view (**b**) masonry (**c**) spillway. Color version available in Appendix

The structures that were constructed during the Hellenistic period could be connected with two unrelated (and probably opposed) to each other characteristics of the Hellenistic world: (a) the growth of the scientific and technological knowledge that permitted more sophisticated solutions in water management, and (b) the turn of the societies to a more luxurious way of life that included the embellishment of the cities and the people thereof. The inventions of Archimedes and Alexandrians and the primitive formulation of fluid mechanics led to the application of advanced hydraulic technology to the Hellenistic cities. Several (inverted) siphons that were parts of aqueducts are the more impressive examples of advanced engineering during this period (Viollet, 2005). The largest of them that belongs to the Pergamon aqueduct and constitutes a miracle of fluid mechanics for that period. It was constructed around 200 B.C. with length that exceeds 3,000 m and it is the first large scale application where water was transported under pressure. On the other hand the raised sanitary and living standards resulted in widespread expansion of private and public baths and lavatories. Their construction was attained to perfection during that period and today their remnants can be found everywhere in Greece (Athens, Delos, Amorgos, Epidaurus, Kos, Philippoi, Thera; Antoniou, 2007). The installation for the cleaning of the young men, at the Gymnasium of Priene (Fig. 6.5) is a well preserved structure of this period.

Fig. 6.5 Installation for the cleaning of the young men, at the Hellenistic Gymnasium of Priene (drawing by V. Kordoni)

The Roman period is characterized by the larger scale of hydraulic works (mainly water supply systems) that include many admirable components. The Roman engineers incorporated the technologies of the previous periods to their experience, and constructed advanced structures such as water bridges, siphons, tunnels, cisterns and urban distribution pipes for the water supply of the cities of Roman Empire. The large number of Roman aqueducts situated everywhere in Mediterranean basin are benchmarks of this ancient technology. Parts of them are preserved in good condition in several sites in Greece (Athens, Corinth, Lappa, Lictos, Mytilene, Nicopolis, Olynthus, Palaiopolis, Patras, Strymi). Finally, after 3000 years of activity, during Roman period the hydraulic knowledge integrated, the structures were spread in every site of the empire and the society was familiarized with the sustainable use of water. In Fig. 6.6, the plan of a Roman residence in the city of Ancient Thera, is presented. Inside the residence there is lavatory and the impluvium a sunken part of the atrium that captured rainwater from the roof. In the same site, a roofed cistern that still stores water is presented in Fig. 6.7.

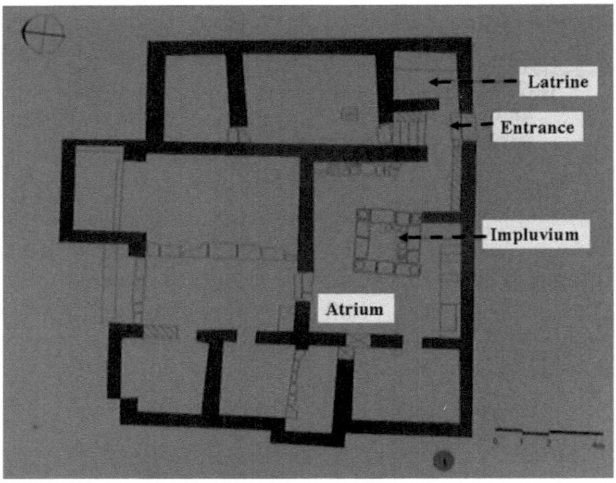

Fig. 6.6 Plan of a Roman residence (city of Ancient Thera). The sites of lavatory and impluvium are visible

Fig. 6.7 Roofed cistern (city of Ancient Thera). Color version available in Appendix

6.4 Discussion

The review of all this information about hydraulic works in ancient Greece reveals several issues that can be also approached from different, non-technical, points of view. Three issues are chosen for further discussion:

a. the cooperation of small- and large-scale structures as a hydrosystem;
b. the relation of the implemented technology with socio-economical characteristics of the societies; and
c. the small-scale facilities that improved the quality of life.

Small-scale hydraulic structures existed in all periods of the Greek antiquity. Urban water supply was not based on a central system, as happens today, but on several structures distributed all over the area. In many important sites, all these structures such as wells, cisterns, fountains, small water distribution systems and aqueducts operated as a hydrosystem. From the small hydrosystems that served the water facilities of the Minoan and Mycenaean palaces gradually Greeks passed to the creation of larger hydrosystems that supported the everyday life of the glorious cities of the Classical period. The operation of the new complex hydrosystems was strongly supported by legislation, institutions and public awareness about water. The Minoan palace sites (Knossos, Zakros, Mallia, etc.) and the city of Athens from Archaic to the Roman period are remarkable examples of such distributed small and large hydrosystems.

The size and character of the projects are related to the socio-economical characteristics of the societies. During oligarchic periods, the political situation favoured the construction of large public works. Those aimed to reflect the society's wealth or the power of the sovereign. For example in the case of the Eupalinos tunnel, even though cheaper works could be easily constructed to serve the same purpose, this breakthrough and expensive engineering solution was chosen perhaps due to an ambition of the local tyrant Polycrates to create a monument of technology (Koutsoyiannis et al., 2008). On the other hand, during the period of democracy in Athens, smaller constructions were preferred, such as cisterns collecting storm water. An advanced institutional framework for the water use that engaged the citizens to sustainable water management, was an important part of the hydrosystem management.

Several small scale facilities, related to water use, improved the quality of life in antiquity. In many cases, these can easily be compared with the modern ones. Lavatories, bathtubs and recreational fountains have been in use since the Minoan period (e.g. the toilets of the Knossos palace had seats and flushing equipment, and were connected to sewers; Angelakis et al., 2005) but they were shaped into an advanced form during the Hellenistic period.

6.5 Summary and Conclusion

Recently ancient water technologies and management practices are being revisited with an increased interest. Motivated from this, an information system is developed to support the scientific research about ancient Greek engineering practices and to disseminate this knowledge to the public.

A quick view of the gathered information reveals that ancient Greeks effectively tackled several water problems that modern societies still have to face up. The lower level of technological means (construction equipment in particular), in comparison to modern standards, did not hinder the design of marvellous hydrosystems with long duration structures that we encounter today everywhere in Greece. Studying these systems of that distant era is worth even today to discover (a) the sustainability that characterizes several management practices and hydraulic works (some of the latter are still functioning to date), (b) the complementary character of projects with different types and sizes, and their relation to the specific socio-economical conditions and (c) the design principles of the engineering solutions that had been applied.

Up to this moment the information system contains data about 100 important hydraulic works from the Minoan era up to the Roman period. It is scheduled that the database will be completed in the future. Also a classification of water management practices and hydraulic devices will be included. Finally, the information will be expanded in space (to other parts of the Ancient Greek world) and time (through the Byzantine period).

References

Angelakis AN, Koutsoyiannis D, Tchobanoglous G (2005) Urban wastewater and stormwater technologies in ancient Greece. Water Res 39(1):210–220

Antoniou GP (2007) Lavatories in ancient Greece. Water Sci Technol 7(1):155–164

Koutsoyiannis D, Mamassis N, Tegos A (2007) Logical and illogical exegeses of hydrometeorological phenomena in ancient Greece. Water Sci Technol 7(1):13–22

Koutsoyiannis D, Zarkadoulas N, Angelakis AN, Tchobanoglous G (2008) Urban water management in ancient Greece: Legacies and lessons. J Water Res Pl-ASCE, 134(1):45–54

Mays LW, Koutsoyiannis D, Angelakis AN (2007) A brief history of urban water supply in antiquity. Water Sci Technol 7(1):1–12

Theodulu T, Memos C (2007) A voyage to ancient Greek harbours on board Limenoscope. Water Sci Technol 7(1):253–260

Viollet PL (2005) Water Engineering in Ancient Civilizations, 5000 Years of History, IAHR

Chapter 7
A Brief History of Roman Water Technology

Larry W. Mays

7.1 Introduction

Rome's power extended only to northern Italy in 265 B.C., where the Celt's were a major barrier to further expansion. They had no navy so that further expansion seemed unlikely; however in the next 120 years Rome became a major Mediterranean power with interests reaching west to Spain and east to Asia and Aegean. In consolidating their empire, the Romans engaged in extensive building of cities. Rome resulted from centuries of irregular growth with particular temple and public districts that were highly planned. The Roman military and colonial towns were laid out in a variation of the grid. The layout of London, Paris and many other European cities resulted from these origins. Cities and towns need a healthy and adequate water supply, so the Romans located along rivers and streams and/or locations with access to springs, which were always favored. When cities were small, obtaining clean water and disposing of wastes was not a major problem, however, as cities grew the large populations and higher densities required public infrastructure.

Historically, settlements and communities relied on natural sources to obtain water. Supplying large quantities of water such as for fountains was a luxury few communities and states could afford before the Roman era. Before water supplies were made available through aqueducts and conduits, many Roman towns relied upon rainwater collection and cisterns for storage. The cisterns ranged from individual use for houses to communal cisterns. Probably the most impressive and immense of the Roman cisterns was the Piscina Mirabilis near Pozzuoli in the bay of Naples, Italy. The early Romans devoted much of their time to public works projects such as building temples, forums, arenas, baths, aqueducts, and sewer systems. The early Roman bourgeois typically had a house of several rooms with a square hole in the roof to let rainfall in and a cistern beneath the roof to store the water.

Romans were more pragmatic than their Greek predecessors in the manner that they planned and constructed the water supply systems. They built what can be

L.W. Mays (✉)
School of Sustainable Engineering and the Built Environment, Arizona State University, Tempe, AZ, 85287-5306, USA
e-mail: mays@asu.edu

L.W. Mays (ed.), *Ancient Water Technologies*, DOI 10.1007/978-90-481-8632-7_7,

called mega water supply systems including many magnificent structures. Sources of water for the Greeks were predominantly groundwater as they avoided surface water, probably because of hygienic reasons. The magnificent aqueducts built by the Romans were supplied by either surface water or groundwater.

7.2 Water Sources and Transmission

7.2.1 Sources

Springs, by far, were the most common sources of water for aqueducts. Water sources for the Roman systems included not only springs, percolation wells, and weirs on streams, but also reservoirs that were developed by building dams. One of the Roman systems that I classify as a mega system was ancient *Augusta Emerita*, at present day Merida, Spain. This system included two reservoirs with dams, now known as the Cornalvo dam and the Proserpina dam. The Proserpina dam is an earthen dam, approximately 427 m long and 12 m high, shown in Fig. 7.1a. The Cornalvo dam is an earthen dam, approximately 194 m long and 20 m high with an 8 m dam crest width, shown in Fig. 7.1b. Both of these dams are still used in the present day, obviously with modifications over the years. Dams were built in many regions of the Roman Empire. Schnitter (1994) and Smith (1971) provide additional information on the history of dams. The source of water for the aqueduct in Segovia, Spain was a weir across the Rio Acebeda.

(a) (b)

Fig. 7.1 Roman dams near Merida, Spain. (Copyright permission with L. W. Mays). Color version available in Appendix

7.2.2 Aqueduct System

Aqueducts were used to transport water from the source to the locations where the water was needed, either for irrigation or for urban water supplies. Roman aqueducts were built to promote quality of life and in general were built to serve existing urban centers, in many cases where prosperous life had existed for centuries before the aqueducts had been built. The Romans built aqueducts largely to supply baths and were an expression of civic pride. Many of the Roman inhabitants obtained their drinking water from wells prior to the construction of the aqueducts. The location and delivery point of the aqueducts were determined by geographical, economic, and social factors.

Roman aqueducts included many components including open channels and pipes. Figure 7.2 shows an example aqueduct system and the downstream urban water distribution system. Obviously there are many system configurations that were built by the Romans, however, the drawing presents the major components including the siphon (inverted siphon) which was used in some systems. Various types of pipes constructed by the Romans included terracotta, lead, wood, and stone. Hodge (2002) provides an excellent discussion of the various types of pipes. Romans used three main types of conduits: (a) open channels (*rivi per canales structiles*), (b) lead pipes (*fistuli plumbei*) and (c) earthenware (terracotta) pipes (*tubili fictiles*). Open channels were built using masonry or were cut in the rock and flows were driven by gravity, while the lead pipes were used for pressurized conduits including inverted siphons.

Figure 7.2 shows the components of an aqueduct system, starting with the source and ending with the distribution basin (*castellum divorsium*), then an urban water distribution system is shown. The many components included covered channels, storage and settling basins, aqueduct bridges, subterranean conduits, arcades of elevated channels. Some aqueducts also included inverted siphons which also included a header tank, a pressurized conduit, a venter bridge, and a receiving tank. Many aqueducts were built below the natural ground level, consisiting of long subterranean conduits. Secondary lines (*vamus*) were built at some locations along the aqueduct to supply additional water. Also subsidiary or branch lines (*ramus*) were used. At distribution points water was delivered through pipes (*fistulae*) made of tile or lead.

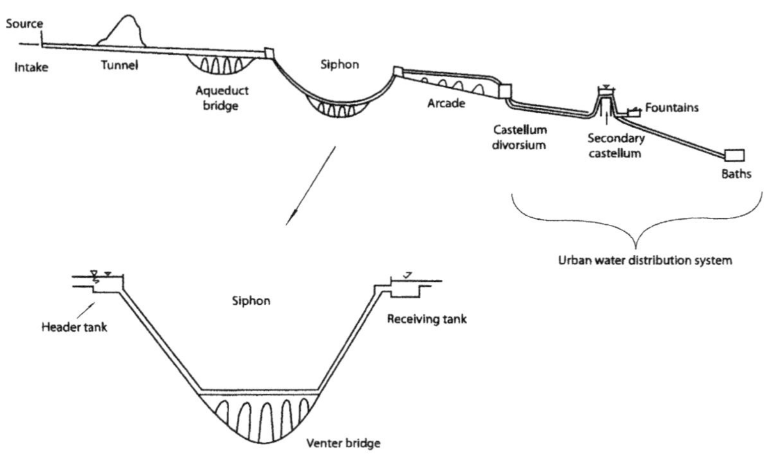

Fig. 7.2 Water supply system showing aqueduct and various components

Fahlbusch (2006) points out the following from examination of several aqueducts:

(a) size of the aqueduct canal was chosen according to the estimated discharge and the size varied along the course of the aqueduct;
(b) the cross-section was large enough for people to walk through the canal for repair and maintenance, particularly to remove calcareous deposits; and
(c) the cross-section was kept constant allowing manifold uses for encasings, especially the soffit scaffoldings for the vaults in a kind of industrialized construction.

There are many examples of Roman aqueducts that could be discussed. Figure 7.3 shows a few examples of Roman aqueducts. One very interesting example (Fig. 7.3b) is the aqueduct of Nemausus (built around 20 B.C.) conveyed water approximately 50 km from Uzes to the *castellum* in the Roman city of Nemausus (present day Nimes, France). From an engineering viewpoint this aqueduct was a remarkable construction project. The elevation difference over the length of the aqueduct was only 17 m, with an average slope of 0.0008.5 m/m and the smallest

(a) (b)

(c) (d)

Fig. 7.3 Examples of Roman aqueducts (**a**) Aqueduct of Gorze near Metz, France. (**b**) Aqueduct of Nemausus just upstream of the Pont du Gard aqueduct bridge. (**c**) Aqueduct of the River Gier near Chaponost, France that supplied the Roman city of Lugdunum (Lyon, France). Color version available in Appendix. (**d**) Pena Cortada aqueduct near Chelva, Spain. Color version available in Appendix. (Copyright permission with L.W. Mays) slope being 0.00007 m/m. The Pont du Gard, one of the most spectacular aqueduct bridges ever built, was part of the aqueduct system to provide water to Nemausus.

Aqueduct bridges were probably the most spectacular feature of the aqueduct systems. One of the most magnificent aqueduct bridges was the Pont du Gard (Bridge of the Gard) shown in Fig. 7.4a. This is the largest of all the Roman aqueduct bridges being 275 m long, 48.4 m high, and a maximum free span of 24.5 m. Figures 7.4b and c show the Roman aqueduct bridge, near Tarragona (Tarraco), Spain, which was part of the Francoli Aqueduct to Tarraco.

Energy dissipation was necessary for aqueduct sections of rather steep slopes even up to 78% (Ashby, 1935; Hodge, 2002), to maintain normal downstream flows and prevent scour and damage to the aqueduct. The structures were (a) smooth steep chutes followed by hydraulic dissipation, (b) stepped chutes, and (c) drop shafts or drop shaft cascades (Chanson 2000, 2002b, 2008).

In some cases instream flow regulation was used (Bossy et al., 2000; Fabre et al., 2000; Chanson, 2002a), that consisted of large basins equipped with sluice gates for regulation. Regulation, most likely, was used to prevent overflows, for maintenance and to optimize flow conditions. As pointed out by Chanson (2008), the storage capacity of aqueducts could be significant (20,000 m^3 for Gorze and 50,000 m^3 for Nimes).

Fig. 7.4 Example Roman aqueduct bridges. (**a**) Pont du Gard aqueduct bridge near Nimes, France. (**b**) Aqueduct bridge of Ferreras (known as Puento del Diablo (devil's bridge) near Tarragona, Spain built during the 1st century B.C. Aqueduct bridge is 217 m long and 27 m high. Bridge consists of two tiers of arches with diameters of 5.9 m (20 Roman feet) with a 15 cm variation and a distance between centers of piers of 7.95 feet (26 Roman feet). (**c**) The *specus* of the aqueduct bridge of Ferreras has a modern day width of 2.5 Roman feet. (Copyright permission with L.W. Mays). Color version available in Appendix

7.2.3 Siphons

The Romans used siphons throughout the empire, and were especially numerous in Gaul, but relatively rare in Rome (Hodge, 2002). The siphons included a header tank for transitioning the open channel flow of the aqueduct into one or more pipes, the bends called *geniculus*, the *venter* bridge to support the pipes in the valley, and the transition of pipe flow to open channel flow using a receiving tank.

One of the largest siphons was the Beaunant siphon of the aqueduct of the Gier River which supplied the Roman city of Lugdunum (Lyon, France). Figure 7.5 shows the siphon ramp (with nine lead pipes), the header tank, and the *venter* bridge. This siphon was 2,600 m long and 123 m deep with an estimated (Hodge, 2002) discharge of 25,000 m^3/24 hour. Total length of the nine parallel pipes was 2.6 km. The largest siphon was built at Pergamon (Madradag) which was 3,000 m long and 190 m deep and only had one pipe compared to the 9 pipes at Beaunant.

(a) (b) (c)

Fig. 7.5 Siphon of aqueduct of Gier, Beaunant, France near Lyon (Ancient Lugdunum). (**a**) This reconstruction shows ramp of siphon with header tank on the top and the nine lead pipes of the siphon (Haberey, 1972 based upon Montauzean, 1908). (**b**) Ramp of the siphon near Chaponost. (Copyright permission with L.W. Mays) (**c**) The *venter* bridge of the siphon near Beaumont. (Photo courtesy of with H. Fahlbusch)

7.3 Vitruvius and Frontinus

To properly discuss Roman water supply we must be aware of the treatises of Vitruvius Pollio, *De Architectura* (Morgan, 1914), and Sextus Julius Frontinus, *De aquaeductu urbis Romae* (C. Hershel translation 1973). Vitruvius (84 B.C.) and Frontinus (A.D. 35–103) did not contribute to the scientific development of hydraulics: however the treatises that they authored do give us insight into the planning, construction, operation, and management of Roman hydraulic systems. The following quote from Vitruvius describes how the aqueduct *castellum* worked (as presented in Evans, 1994). "When it (the water) has reached the city, build a reservoir with a distribution tank in three compartments connected with reservoir to receive the water, and let the reservoir have three pipes, one for each of the connecting tanks, so that when the water runs over from the tanks at the ends, it may run into the one between them." Frontinus was a retired army officer who, in A.D. 97 took over as the director of the Roman Metropolitan Waterworks. He declared the Roman aqueduct as the real mark of civilized living. Frontinus is discussed further in Chapter 11.

Frontinus provided measures of volume of the aqueducts of Rome in *quinaria*, which unfortunately, cannot be converted into modern units of measurement (Evans, 1994), as we do not know the definition of the term. For the seven aqueducts of Rome that existed during his time, he reported they had a volume of 9,955 *quinaria*.

7.4 Urban Water Distribution Systems

7.4.1 Water Distribution

There were many components to the Roman urban water distribution system. Figure 7.2 illustrates how the water distribution system connects to the aqueduct system. Water flowed by gravity through enclosed conduits (*specus* or *rivus*), which typically were underground, from the source to a terminus or distribution tank (*castellum*). Above ground aqueducts were built on a raised embankment (*substructio*) or on an arcade or bridge. Settling tanks (*piscinae*) were located along the aqueducts to remove sediments and foreign matter. These pipes were connected to the *castellum* by a fitting or nozzle (*calix*). These pipes were placed below ground level along major streets.

Pompeii (located on the Bay of Naples, south-southeast of Mt. Vesuvius in Italy) had a water supply system that is representative of Roman urban water distribution systems. Sources of water for Pompeii included wells, cisterns and other reservoirs, and the Serino aqueduct. This aqueduct received water from springs at Serino, near Avellino, and then was routed via Sarno around the north side of Mt. Vesuvius to serve Naples and two large cisterns of Cento Camerelle (Baiae) and the Piscina Mirabilis (Misenum). From Sarno a branch aqueduct was routed to Pompeii terminating at the *castellum* at Porta Vesuvii, shown in Fig. 7.6. According to Richardson (1988, p. 51) there were no springs within the city of Pompeii. The water table was tapped within Pompeii using wells as deep as 38 m below the surface (Maiuri, 1931, pp. 546–557).

Two approaches to laying out the pipe network were followed: (a) using a main pipe from the secondary *castellum* with smaller branch pipes attached to serve individual customers, and (b) not using main pipes but using individual pipes laid from the secondary *castellum* to the individual customer, which was the normal Roman practice (Hodge, 2002). Pompeii's water distribution system consisted of pipes along the main streets connecting the main *castellum* at Porta Vesuvii (Fig. 7.6a) to the various water towers (secondary *castella*) (Fig. 7.7a), from which smaller pipes (Fig. 7.7b) were placed under the sidewalks and streets and served the various customers. Not all customers had individual lines from a secondary *castellum* but instead received their supply from taps into the system at their locations.

A water tower (secondary *castella*) that had a lead storage tank placed on top of the tower with a public water fountain on the street is shown in Fig. 7.7a. Public water fountains, supplied from the secondary *castella*, were placed somewhat evenly around the urban center, being about 50 m apart. The fountains had an overflow weir so that the water would flow onto the streets and then into the drainage system. Terracotta pipes were not used in the water distribution system in Pompeii (Jansen, 2001). Lead pipes (Fig. 7.7b) in Pompeii are of the same construction and appearance as found in other Roman cities. The water taps found in Pompeii were also similar to those found in other Roman cities. Only a small number of houses had a water pipe that supplied a private bath or basins in the kitchen. Overflows were drained into cisterns for rainwater.

The households and public buildings both had very interesting systems to collect and store rainwater. Buildings with peaked roofs had gutters along the eaves to collect the rainwater and downspouts to carry the water to the cisterns located under the building. Downspouts were made of terracotta pipes and were often set inside the wall.

Fig. 7.6 *Castellum divisorium* in Pompei. (**a**) Building housing the *castellum*. Color version available in Appendix. (**b–c**) Three gated channels of the *castellum* used to control flow to geographical areas of Pompeii. (Copyright permission with L.W. Mays)

Fig. 7.7 Components of water distribution system in Pompei. Color version available in Appendix. (a) Water tower (secondary *castella*) with a storage tank was mounted on top of the tower. Lead pipes were used for the flow of water to and from the tank and were placed in the vertical recessed portion on the tower. The exiting pipes (*calices*) branched off to supply various customers and public fountains. (b) Lead pipe and joint found along the street. (c) Junction box. (Copyright permission with L.W. Mays)

7.4.2 Cisterns

The Romans made extensive use of cisterns, so that herein I will only be able to explore a very few of the many that were built. Cisterns were used extensively for storing water from rainfall collection and from aqueducts. One example of a cistern (Fig. 7.8a) is at the Roman city of Ilici (La Alcudia de Elche) near present day Elchi, Spain. Figure 7.8b shows a Roman cistern at the base of the Acropolis in Athens, Greece. In the Roman town of Pompeii, with the extensive water distribution system including both aqueduct water and well water, the roofs of houses collected rainwater that flowed through terracotta pipes down to cisterns where water was stored for domestic use.

The Piscina Mirabilis, near Naples, Italy, is one of the largest Roman cisterns (capacity of 12,600 m^3 of water). The cistern was supplied by water from the Augustan aqueduct, the Serino aqueduct that was built from Serino to Miseno. The Serino aqueduct, 96 km long with seven branches, supplied many towns including Pompeii, Herculaneum, Acerra, Atella, Nola, and others. The total elevation drop in elevation from the source, the Acquaro-Pelosi spring in Serino to the Piscina

(a) (b)

Fig. 7.8 Roman Cisterns (**a**) Cistern in Illici, Spain (**b**) Cistern below Acropolis in Athens. (Copyright permission with L.W. Mays). Color version available in Appendix

Mirabilis is 366 m (0.38%). This large cistern is 72 m by 27 m in plan and is 15 m deep (according to Hodge, 2002).

In Roman North Africa vast cistern complexes were used in conjunction with the aqueducts (Wilson, 2001). These cisterns had capacities that were often several thousand m^3, that were much larger than the domestic cisterns. They were typically located where the aqueducts reached the edge of towns. Wilson (2001) describes two types of common cistern complexes in North Africa, both of which were used at Uthina in Oudna, Tunsia. Large cross-vaulted chambers, with a roof supported by piers, is one type of cistern. A second common type of cistern complex includes several barrel-vaulted chambers with a transverse chamber set across them.

In the cisterns at Tuccabor and Djebel M'rabba in Tunsia, the transverse chamber was placed between the inlet and the parallel chambers and the chamber serves as a settling tank before water enters the storage chambers (Wilson, 2001). At Thugga, Thuburnica, Thapsus and Uthina (Fig. 7.9), the transverse chamber is placed between the parallel chambers and the outlet, with no settling (Wilson, 2001). At Thuburnica and the Ain el-Hammam cisterns at Thugga the entrance of the aqueduct channel runs along an internal wall of the cistern so that it distributes water to the cistern chambers. Cisterns at Dar Saniat at Carthage were constructed with three settling basins and two storage reservoirs each of two compartments with a total storage capacity of 2,780 m^3 (see Fig. 7.10). Primary settling tank A (oval in shape) received water from the aqueduct and water entered the two–chamber cistern (D and E). Water also flowed from settling tank A into secondary circular settling tanks B and C before entering the second cistern chambers F and G. The water in F and G obviously would have been cleaner. A circular tap chamber (H in Fig. 7.10) received water through two lead pipes from D and E at floor level. It also received the higher quality water from G and F in a third lead pipe a meter higher than floor level.

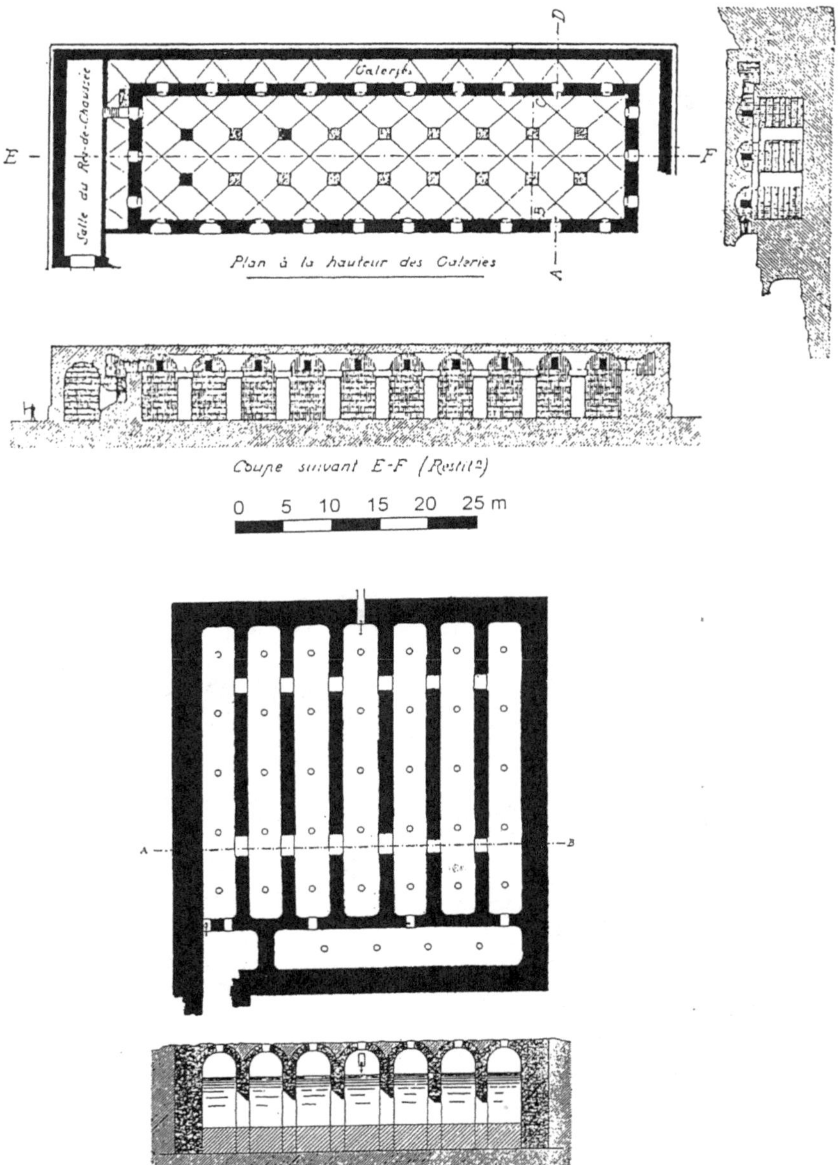

Fig. 7.9 Plans and sections of the small (**top**) and large (**bottom**) cisterns at Uthina (Oudna, Tunsia) (after Babelon and Cagnat 1893, text to f.XXVIII, Oudna; as presented in Wilson (2001))

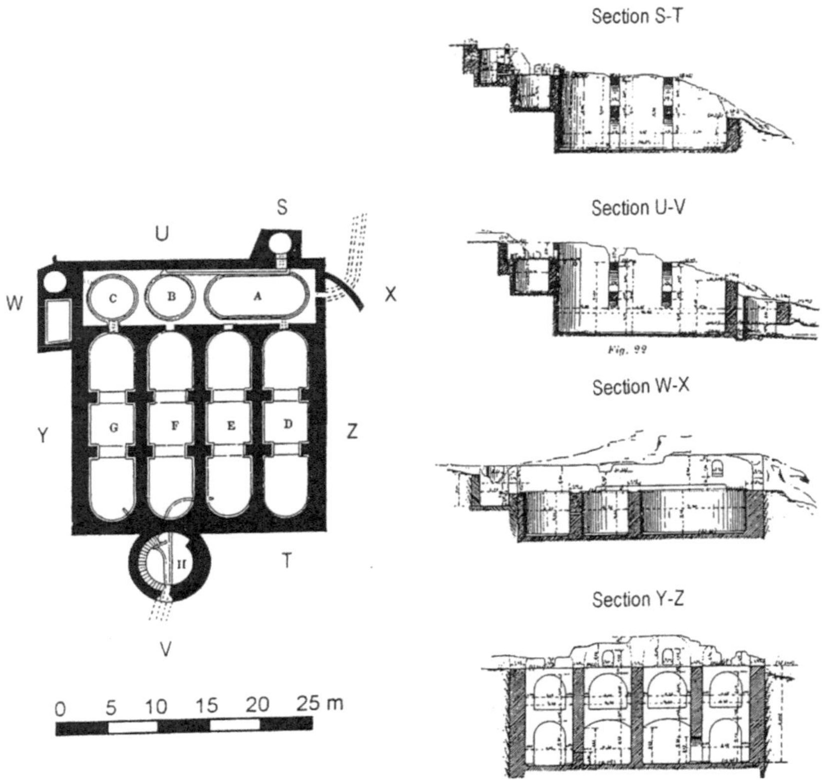

Fig. 7.10 Cisterns at Dar Saniat at Carthage (after Cagnat and Chapot, 1916, 1:89, Fig. 43, and Renault, 1912, 478–479, as presented in Wilson (2001))

The story of cisterns continues to the magnificent cisterns built by the Byzantines. These include one of the largest and most magnificent cisterns of all, the Yerebatan Saryi or Basilica Cistern in Istanbul, Turkey. Other great cisterns and even cistern systems were built. In Aden, Yemen, built on the cone of a huge extinct volcano, tunnels and channels were built to transport water to a series of 50 large open air cisterns. Even today the use of cisterns remains a very important aspect of water supply in many parts of the world.

7.4.3 Fountains, Ornamental Pools, and Baths

Above the public fountains in Pompeii are briefly discussed. Now we briefly discuss the more monumental fountains (*nymphaeum*) and use as an example the Roman fountain found in Perge, approximately 20 km east of the present day city of Anatalya, Turkey. Figure 7.11a–c illustrates the magnificent fountain that was built in Perge. Figure 7.11c shows the drain opening in canal downstream of fountain. Figure 7.12 shows the monumental fountain (*nymphaeum*) at Miletus (Turkey) of

Fig. 7.11 Perge, Turkey (**a**) Two-story monumental fountain where water overflowed into a canal that divided the colonnaded street. Covered canal delivered water to a pool behind the façade. Water from the pool flowed through an opening just below the reclining statue of the river god Cestrus. (**a**) Fountain and the canal downstream which divided the colonnaded street. Color version available in Appendix (**b**) Close-up view of fountain showing Cestrus. (**c**) Drain opening in canal downstream of fountain. (Copyright permission with L.W. Mays)

Fig. 7.12 Remains of monumental fountain in Miletus (Turkey). To the left in back of the fountain is part of the Roman aqueduct that supplied water to the fountain. (Copyright permission with L.W. Mays). Color version available in Appendix

Fig. 7.13 Fountain in ancient Chersonisos, Crete, Greece. Mosaic shows a fisherman and many fish. (Copyright permission with L.W. Mays). Color version available in Appendix

ancient Ionia, with the aqueduct in the back that supplied the fountain with water. The fountain had three stories, 17 m high and 20 m wide, with a large pool in the front. Today as illustrated in the photo is basically destroyed, with only fragments of the first floor with water tanks consisting of niches in front and three partitions in the rear remaining. All floors of the marbled-faced façade had Corinthian columns, where statues of gods, goddesses and nymphs stood in the niches between the columns. Water flowed from amphorae (large earthernware storage vessels with an oval body and two handles extending from below the lip to the shoulder) held by the statues on the first floor into the pool. On the second floor water flowed from the mouths of fish into the pools. Indications are that the first two floors of the fountain were built in two different time periods. On the first floor is the inscription naming Emperor Trajan (98–117 A.D.) and on the second floor is the inscription of Emperor Gordian III (214–244 A.D.).

Figure 7.13 shows the Roman fountain in ancient Chersonisos located near the present harbor or Limani Chersonisos, Crete, Greece. Note the fish and fisherman in the fountain. The water supply system of Chersonisos in Roman times included wells, springs, and an aqueduct. The source of the aqueduct has not always been clear. The terminus of the aqueduct is a reservoir in Palatia village. Figure 7.14 shows an ornamental pool and channel, the canopus at Villa Hadrian's Villa (Villa Adriano) at Tivoli, Italy.

Baths were unique in ancient Roman cities. One example is the Skolacctia baths in Ephesus, which had a salon and central heating. Baths had a hot bath (*caldarium*), a warm bath (*tepidarium*), a cold bath (*frigidarium*), and a dressing room (*apodyterium*). The baths would have many public and private rooms. A furnace and a large boiler were used to provide hot water. Figure 7.15 shows examples of Roman baths. There are many other examples of baths throughout the Roman Empire.

Fig. 7.14 Canopus at Hadrian's Villa (Villa Adriano) at Tivoli, Italy. (Copyright permission with L.W. Mays). Color version available in Appendix

Fig. 7.15 Roman Baths (**a**) and (**b**) Baths at Perge, Turkey. (**c**) Reconstruction of bath at Lucentum (present day Alicante, Spain) (**d**) Bath at Hadrian's Villa at Tivoli, Italy. (Copyright permission with L.W. Mays)

7.5 Roman Mega Water Projects

The Romans built what can be called mega water supply systems including many magnificent structures. The water system of aqueducts and dams to Merida, Spain (ancient Augusta Emerita) is described in this section. Three of the magnificent mega water systems that are not described herein include: the system of aqueducts with various structures to Lyon, France (ancient *Lugdunum*); the system of aqueducts to Rome, and the aqueduct of Nimes (ancient *Nemausus*). There are many others that include the components of the Roman systems. See Hodge (2002), Mays (2002) and Mays et al. (2007), among many other references for details of these systems.

In 25 B.C. Emerita Augusta (present day Merida, Spain) became a colony and a century later the Romans had built a water supply system including three aqueducts (Fig. 7.16), two of which were supplied by dams (the Proserpina dam, Fig. 7.1a, and the Cornalvo dam, Fig. 7.1b). The three aqueducts were the Cornalvo aqueduct (enters on the east side of Merida), the Proserpina aqueduct (enters on the northwest side of town), and the Las Thomas aqueduct (from springs on the north and northeast side of Merida). The Cornalvo aqueduct was built first and was about 17 km long. Cornalvo dam (Fig. 7.1b) is an earthen dam approximately 194 m long, 20 m high, and has an 8 m dam crest width. A few remains of the aqueduct are visible (Fig. 7.17e) near the present day bull ring. The Las Thomas aqueduct included an

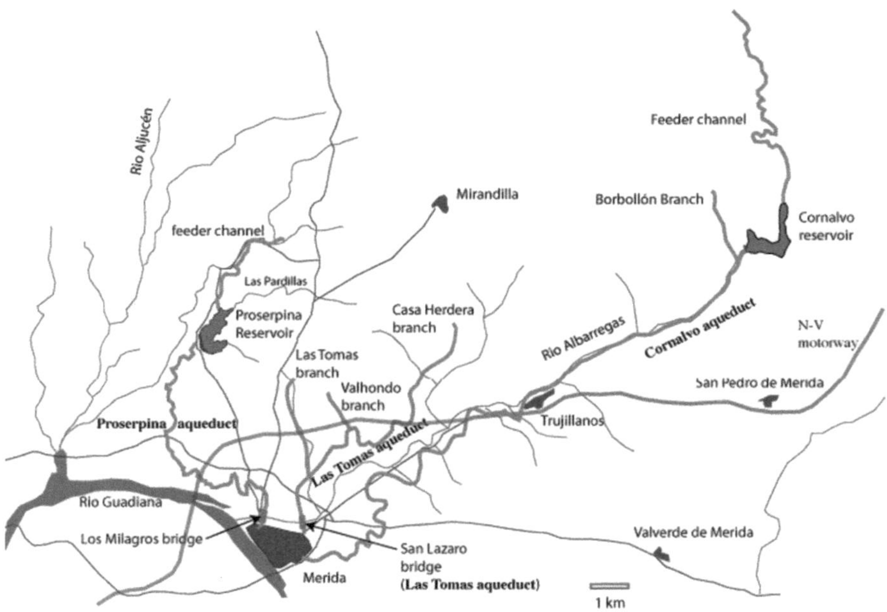

Fig. 7.16 Map showing the three Roman aqueducts in Merida, Spain. (Copyright permission with C.W. Passchier of Mainz, Germany, used with permission)

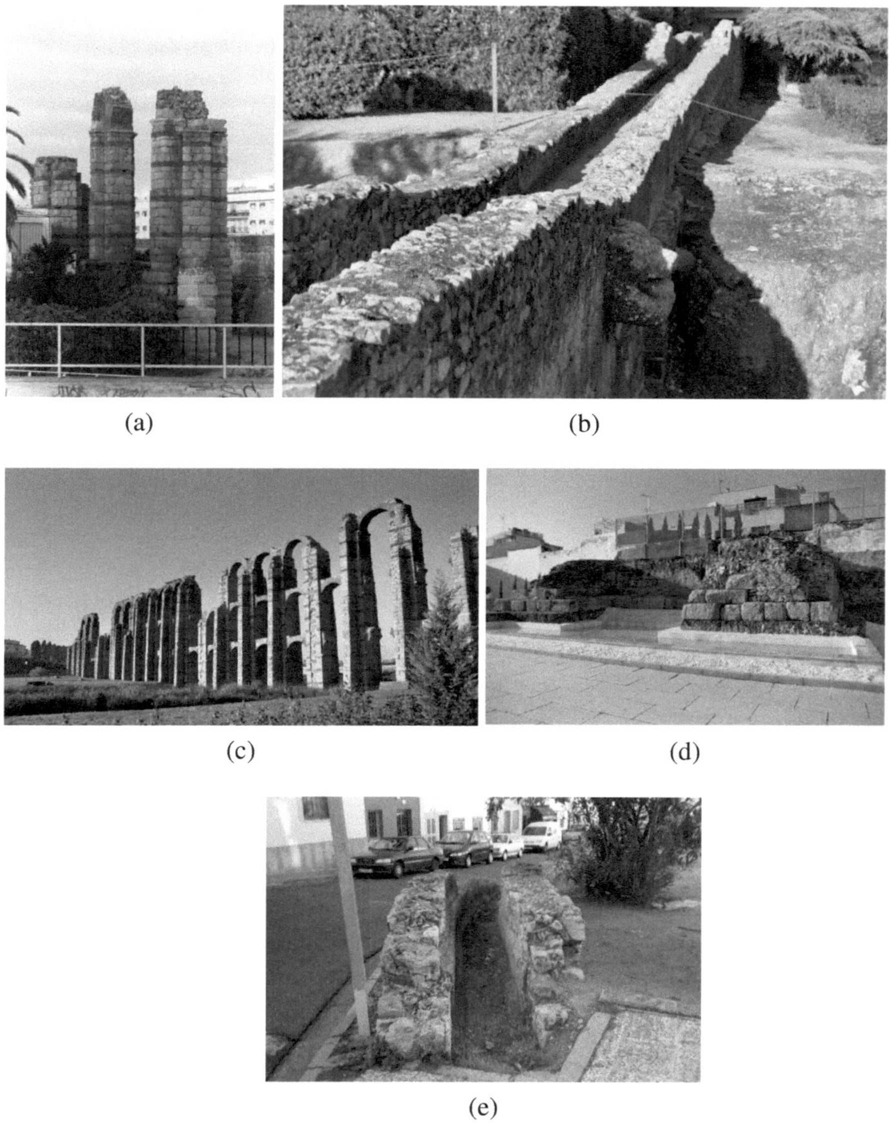

Fig. 7.17 Components of the water supply system in Merida (**a**) Only remains of the Las Thomas aqueduct bridge across the Rio Albarregas located near the hippodrome; (**b**) Las Thomas aqueduct downstream of the aqueduct bridge near the Roman theater. Shown on the right side of the aqueduct is a lion head made of stone which was used as a gutter spout. The remains of a castellum is nearby. (**c**) Los Milagros aqueduct bridge of the Proserpina aqueduct across the Rio Albarrregas; (**d**) The remains and the reconstruction of the castellum downstream of the Los Milagros bridge; (**e**) Cornalvo aqueduct supplying water from the Cornalvo dam to Merida. Only a few remants of this aqueduct are still visible near the bullfighting arena. (Copyright permission with L.W. Mays). Color version available in Appendix

aqueduct bridge 1,600 m long (across the Rio Albarregas), of which only three pillars (16 m high) remain (see Fig. 7.17a). Materials from this aqueduct bridge were used by the Arabs in the 16th century to construct the San Lazaro aqueduct bridge. The Proserpina dam (see Fig. 7.1a) is an earthen dam, 427 m long and 12 m high, is located north of Merida. The Proserpina aqueduct was about 10 km long, starting at the Proserpina dam, and entered the town on the north side with an aqueduct bridge over the Rio Albarregas. This aqueduct bridge (see Fig. 7.17c), referred to as the Los Milagros (the miracles) by the Spanish, has a maximum height of 30 m. The aqueduct ends at the remains of the castellum divisorium (see Fig. 7.17d).

Four aqueducts were used to supply water to the ancient city of Lugdunum (Lyon, France). They were the Mont d'Or, the Yzeron, the Brevenne, and the Gier. The aqueduct of the River Gier (Fig. 7.3c), was the longest and the highestof the four aqueducts. Approximately half of the aqueduct was subterranean with at least nine tunnels and four siphons. This aqueduct had four siphons and over 80 manholes.

The aqueduct system in Rome evolved over a 500-year time period, with the first aqueduct, the Aqua Appia, being constructed around 313 B.C. This system eventually consisted of 11 aqueducts that eventually supplied water to Rome from mostly springs, and two were supplied from the Anio River and one from Lake Alsietinus. All the major eastern aqueducts entered Rome at the Porta Maggiore. The aqueducts of Rome are discussed further in Chapter 11.

7.6 Other Types of Water Technology

7.6.1 Roman Hydraulic Devices

The force pump, a water lifting device, was invented by the Greek inventor and mathematician, Ctesibius (or Ktesibios or Tesibius) of Alexandria (ca. 285–222 B.C.). Roman force pumps were more technically complex than other ancient water lifting devices. Two cylinders mounted vertically were connected near their bases by a horizontal pipe with a single upright delivery pipe at the center of the pipe (see Fig. 7.18). Bronze flap valves (intake valves) were hinged at the bottom of each cylinder so they would open inwardly into the cylinder. A bronze piston was placed inside each cylinder so that when it was lifted (the up stroke) from the base of the cylinder, water entered through the flap valve and filled the cylinder. When the piston was force downward (down stroke) water entered the transverse pipe connecting the two cylinders as the intake valve at the base of the cylinder closed. The upright delivery pipe had a one-way flap valve at its base which kept water moving in the vertical direction. The pistons in the two cylinders were moved by means of a connecting rod attached to opposite ends of a single lever. This allowed the pistons to work reciprocally so that when one piston was down the other was up. Figure 7.18a shows the type of force pump described above. The force pump in Fig. 7.18b was constructed using beveled spindle valves instead of flap valves. Water was discharged from the force pump through a pipe or nozzle.

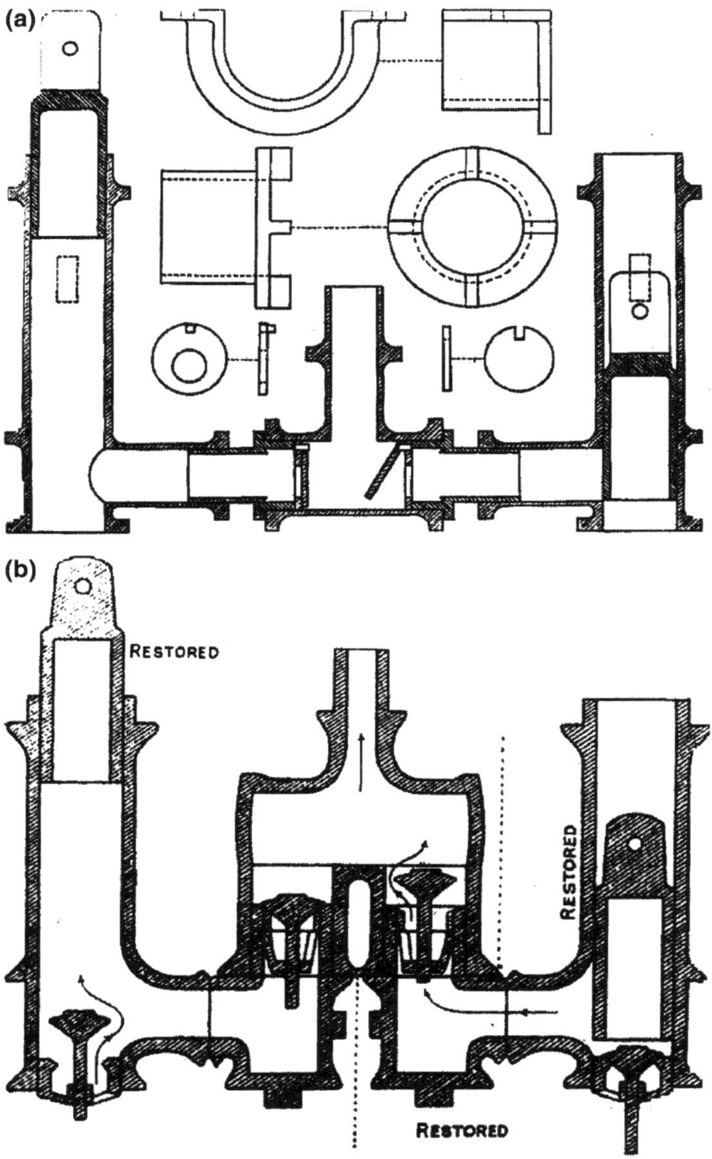

Fig. 7.18 Reconstruction of force pumps found at Bolsena, Italy (as presented in Oleson, 1984). (**a**) F. Davis, Note on a Roman Force Pump Found at Bolsena, *Archaeologia* 5, 1896. (**b**) H.B. Walters, A Guide to the Exhibition Illustrating Greek and Roman Life, 3rd ed. London: British Museum, 1929

7.6.2 Water Mills

Water mills were another important hydraulic device used in antiquity (Wikander, 2000). Two main types of water mills were developed according to the position of the water wheel, vertical and horizontal. Vertical water mills required a right-angle gear and horizontal water mills had a rotating millstone attached to the vertical shaft. The vertical water wheel turns a horizontal wheel shaft, with a vertical cog-wheel attached at the other end. Three functional variants of the vertical waterwheel existed. An undershot wheel was immersed into a watercourse and powered by running water. The overshot wheel was powered by water conducted through a wooden chute above the wheel. Both weight of the water and impulse helped to turn the

(a) (b) (c)

Fig. 7.19 Roman water mill located in the Greek Agora in Athens. This vertical water mill was an overshot type as evidenced by the way in which the lime deposits formed on the walls. (**a**) Mill race (branch of Hadrians aqueduct) for Roman mill in Athenian Agora. Valerian Wall shown to the right. (**b**) Wheel pit (or wheel race) for water mill with a branch of Hadrian"s aqueduct supplying water for the mill shown in the background in top of picture. The mill race is shown in the top of the picture. Note the carved area on the right where the horizontal wheel shaft was mounted and the opening for the wheel shaft on the left. The wheel shaft was connected to a vertical cogwheel located outside the wheel pit. Color version available in Appendix (**c**) Water flowed from the wheel pit (wheel race) through the drain on the north wall in the bottom of the pit. Also shown on the east wall are the scratches made by the wheel. Color version available in Appendix. (Copyright permission with L.W. Mays) wheel. A breast-shot wheel was powered by water hitting the back of the wheel, which was a middle approach to the undershot and the overshot. The horizontal wheel mill was used in mountainous areas where higher heads and velocities could be used so that water hit oblique paddles (vanes) through a steep chute.

Figure 7.19a–c show the wheel pit (slot like structure set into the bedrock) of a Roman water mill supplied by a branch of Hadrian's aqueduct located in Athens, Greece. The walls of the pit were thickly coated with lime deposit from the hard Athenian water. The mill was located in the Athenian Agora near the Stoa of Attalos by the Valerian Wall (shown in Fig. 7.19a). Parsons (1936) described in detail this particular mill which was uncovered in 1933. The wheel for this mill had a diameter of 3.24 m.

There are also what can be called multiple mills (Hodge, 2002) in which a number of wheels are arranged in series or parallel. One is located in Israel at Nahal Tanninim near Caesarea on the Crocodile River. This facility includes a dam for a reservoir to supply the aqueduct to Caesara where a number of horizontal mills are arranged in parallel. Two of the mills have been identified as Roman and the others Turkish (Hodge, 2002). Another site of multiple mills (three mills in parallel) is located in western Tunsia on the Medjerda River at Chemtou (ancient Simitthus). Barbegal (located 2 km south of Fontvielle, north-east of Arles in southern France) has two parallel rows of mills and supplied by a 9 km long aqueduct. On each row of mills are located eight mill houses, so a total of sixteen single vertical wheels, probably overshot, though the undershot type has been suggested (Hodge, 2002).

7.7 Conclusions

In this chapter the development of water technology by the Romans has been traced. We have looked at some of the major advances that occurred in hydraulic technology during antiquity long before the development of the laws of conservation used in modern day design of hydraulic structures. The Romans did not add to science like the Greeks: however, they contributed tremendously to the advancement of engineering. They also invented new technologies such as the Roman concrete (*opus caementitium*) which allowed the construction of long canals, very large bridges and long tunnels in soft rock (Fahlbusch, 2006). Considering the hydraulic structures that were built by the ancients, they had a remarkable observation of nature and were able to draw conclusions from their observations.

References

Ashby T (1935) The aqueducts of ancient Rome. Clarendon Press, Oxford, UK, edited by IA Richmond

Bossy G, Fabre G, Glard Y, Joseph C (2000) Sur le Fonctionnement d'un Ouvrage de Grande Hydraulique Antique, l'Aqueduc de Nîmes et le Pont du Gard (Languedoc, France). Comptes Rendus de l'Académie des Sciences de Paris, Sciences de la Terre etdes Planètes 330:769–775

Chanson H (2002a) Certains Aspects de la Conception hydrauliques des Aqueducs Romains. (Some Aspect on the Hydraulic Design of Roman Aqueducts.) Jl La Houille Blanche 6/7:43–57
Chanson H (2002b) An experimental study of Roman dropshaft hydraulics. J Hyd Res, IAHR 40(1):3–12
Chanson H (2008) The hydraulics of Roman aqueducts: What we know? Why should we learn? Proceedings World Environmental and Water Resources Congress
Evans HB (1994) Water distribution in ancient Rome: The evidence of Frontius. University of Michigan Press, Ann Arbor, USA
Fabre G, Fiches JL, Paillet JL (2000) L'Aqueduc de Nîmes et le Pont du Gard. Archéologie, Géosystème, Histoire. CNRS Editions, CRA Monographies Hors Série, Paris, France
Fahlbusch H (2006) Water Engineering and Management in the Classic Civilizations, Proceedings, La Ingenieria Y La Gestion Del Agua a Traves de Los Tiempos, held at the Universidad de Alicante, Spain, with the Universidad Politechnica de Valencia, Spain
Frontinus, Sextus Julius, The Two Books on the Water Supply for the City of Rome, A.D. 97, translated by C. Hershel, republished by the New England Water Works Association, Boston, MA, 1973.
Hodge AT (2002) Roman aqueducts and water supply. Gerald Duckworth & Co, Ltd, London, USA
Jansen GCM (2001) Water pipe systems in the houses of Pompeii. In: Koloski-Ostrow AO (eds) Water use and hydraulics in the Roman city. Archaelogical Institute of America, Colloquia and Conference Papers, No. 3, Kendall/Hunt Publishing Company, Dubuque, USA
Maiuri A (1931) Pompeii Pozzi e Condotture d'Acqua... and Scoperta di Grandi Conditture in Piomb dell'Acqueducto Urbano. Notizie degli Scavi di Antichita, 7
Mays LW (2002) Urban water infrastructure: A historical perspective. In: Mays LW (ed) Urban water supply handbook, McGraw-Hill, New York, USA
Mays LW, Koutsoyiannis D, Angelakis AN (2007) A brief history of urban water distribution. Water Sci Technol: Water Supply, IWA Publishing, 7(1):1–12
Montauzan Gde (1908) Les aqueducs antiques des Lyon, Paris
Morgan MH (trans) (1914) Vitruvius, The ten books on architecture. Cambridge, MA
Parsons AW (1936) A Roman water-mill in the Athenian Agora, Hesperia 5:70–90
Richardson L Jr (1988) Pompeii: An architectural history. John Hopkins University Press, Baltimore, USA
Schnitter NJ (1994) A history of dams, the useful pyramids. A. A. Balkema, Rotterdam, Netherlands
Smith N (1971) A history of dams. Peter Davies, London
Wilson A (2001) Urban water storage, distribution, and usage. In: Koloski-Ostrow AO (ed) Water use and hydraulics in the Roman city, Kendall/Hunt Publishing Company, Dubuque, Iowa.

Bibliography

Aicher PJ (1995) Guide to the aqueducts of Rome, Bolchazy-Carducci Publisher's Inc., Wauconda, IL
Amit D, Patrich J, Hirschfeld Y (2002) The aqueducts of Israel. J Roman Arch, Suppl. Series No. 46
Blackman DR (1978) The volume of water delivered by the four great aqueducts of Rome. Papers of the British School at Rome 46:52–72
Blackman DR (1979) The length of the four great aqueducts of Rome. Papers of theBritish School at Rome 47:12–18
Blackman DR, Hodge AT (2001) Frontinus' Legacy. Essays on Frontinus' de aquis urbis Romae. University of Michigan Press, Ann Arbor, USA, 170 pp

Butler S (1774–1839) The atlas of ancient and classical geography, editor Ernest Rhys (1859–1946), www.gutenberg.org/ebooks/17124

Chanson H (2000) Hydraulics of Roman aqueducts: steep chutes, cascades and dropshafts. Am J Archaeol 104(1):47–72

Chanson H (2001) The hydraulics of stepped chutes and spillways. Balkema, Lisse, The Netherlands

Chanson H (2002) Hydraulics of a large culvert beneath the Roman fqueduct of Nîmes. J Irrigation Drainage Eng, ASCE, 128(5):326–330

Chanson H (2004). Hydraulics of rectangular dropshafts. J Irrigation Drainage Eng, ASCE, 130(6):523–529

Chanson H (2007) Air entrainment processes in rectangular dropshafts at large flows. J Hydraulic Res, IAHR, 45(1):42–53

Fabre G, Fiches JL, Leveau P, Paillet JL (1992) The Pont du Gard. Water and the Roman town. Presses du CNRS, Caisse Nationale des Monuments Historiques et des Sites, Collection Patrimoine au Présent, Paris, France

Garbrecht G (1987). Hydraulics and hydraulic research: a historical review. Balkema Publ, Rotterdam, The Netherlands

Haberey W (1972) Die romischen Wasserleitungen nach Koln. Rheinland-Verlag, Bonn

Hauck G (1988) The aqueduct of Nemausus, McFarland & Company, Inc. Jefferson, NC and London

Hodge AT (ed) (1991) Future currents in aqueduct studies, Leeds

Hodge AT (1998) Ancient Greek France, University of Pennsylvania Press, Philadelphia

Koloski-Ostrow AN (ed) (2001) Water use and hydraulics in the Roman city, Colloquia and Conference Papers, Number 3, Archaeological Institute of America, Boston, MA, Kendall/Hunt Publishing Company, Dubuque Iowa

Ozis U (1987) Ancient water works in Anatolia. Water Res Develop 3(1), Butterworth & Co. Ltd.

Ozis U (1996) Historical water schemes in Turkey. Water Res Devel 12(3), Butterworth & Co. Ltd.

Shepard WR (1911) The historical atlas, http://ancienthistory.about.com/od/romaancientRome

Viollet PL (2007) Water engineering in ancient civilizations. 5,000 years of history, IAHR Monograph, Madrid, Spain

Wikander O (2000) Handbook of ancient water technology. Koninklijke Brill NV, Leiden, The Netherlands

Chapter 8
Analysis of the Water System of the Ancient Roman City of Apamea

Benoît Haut and Didier Viviers

8.1 Introduction

In this chapter, a hydraulic system of the city of Apamea (Syria), operated during the 6th century A.D. (Byzantine period), is characterized. Flow rates and energy losses are calculated using global energy balances, coupled with local Computational Fluid Dynamics simulations, realized with the commercial software Fluent 6.3. This characterization provides a new quantitative description of the water supply system of the city, supplementing the usual field observations. Furthermore, the results of this hydraulic characterization are analyzed from an archaeological point of view. Three open historical questions about Apamea are considered. The analysis of the results of the hydraulic characterization allows to propose new interesting elements regarding the answer to these three questions.

8.1.1 A Brief History of Apamea

The archaeological site of Apamea in Syria, on the right bank of the Orontes, between Hama and Aleppo (Fig. 8.1), had a continuous human activity that can be tracked back to the Middle Palaeolithic. After the conquest of Alexander the Great, a Greek city was established there in 300/299 B.C. by Seleucos Nicator, the King of Syria. Apamea became one of the main cities of the Seleucid Empire, with Antioch, Seleucia or Laodicea. Apamea imposed itself as an administrative and military centre of North Syria.

Later on, Apamea was attached by Pompeii to the Roman Empire (64 B.C.). Because of several violent earthquakes, the city was characterized by large reconstructions, thanks to the liberalities of Roman Emperors.

As the Capital of the Roman province of Syria Secunda, with a dense population, Apamea reached genuine prosperity during the 5th and 6th centuries A.D.

B. Haut (✉)
Université Libre de Bruxelles, Transfers, Interfaces and Processes (TIPs) – Chemical Engineering Unit, Av. F.D. Roosevelt 50, C.P. 165/67, 1050 Brussels, Belgium
e-mail: bhaut@ulb.ac.be

L.W. Mays (ed.), *Ancient Water Technologies*, DOI 10.1007/978-90-481-8632-7_8,

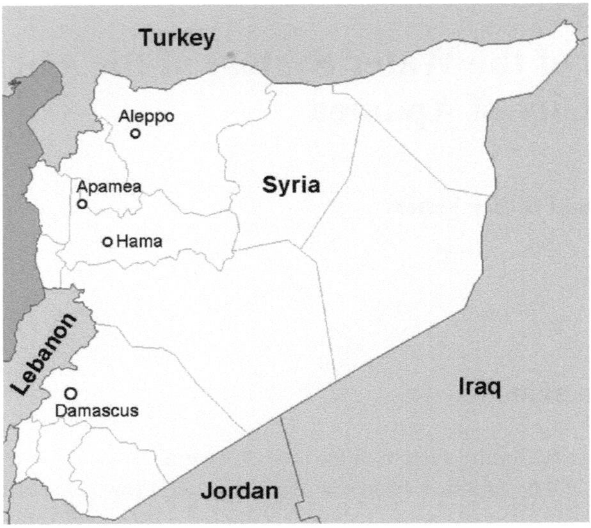

Fig. 8.1 Localisation of Apamea

(the Byzantine period). The Persian Wars combined with dramatic earthquakes subsequently weakened the city and made it vulnerable to the Arabic conquest in 638 A.D.

Since at least the beginning of the second century A.D., a large street with porticoes offered to the city a monumental artery by which the town was intersected from North to South (Fig. 8.2). This Cardo Maximus, approximately 37 m wide and 1850 m long, was also used for wheeled transport during the Roman period. A second feature of urbanism was the large wall that surrounded the city. The foundations of this about 7 km long fortification were set in the Hellenistic times. The only

Fig. 8.2 The Cardo Maximus. Color version available in Appendix

known water supply of the city was an aqueduct, used from 47 A.D. (Balty, 2000) until at least the 7th century. It was bringing water into the town from a spring located about 80 km far from Apamea (Balty, 1987).

8.1.2 Excavations in the North Eastern Area of the City

Excavations in the North Eastern area of the city, where the aqueduct goes into the town, were performed in the last ten years by a team of archaeologists from the Université Libre de Bruxelles. In Fig. 8.3, an overview of these excavations is presented. They reveal at least three main periods of (re)construction, characterized by different water systems (Viviers and Vokaer, 2009).

Fig. 8.3 Excavations in the North Eastern area of the city. Color version available in Appendix

8.1.2.1 Phase 1

The first hydraulic system was used from 47 A.D. to the end of the 4th century A.D. (Viviers and Vokaer, 2009). Only some fragments of this hydraulic system were excavated. In Fig. 8.4, some excavated terracotta pipes used during phase 1 are presented.

8.1.2.2 Phase 2

The second hydraulic system was used from the end of the 4th century A.D. to the end of the 5th century A.D. (Viviers and Vokaer, 2009).

The flow of water in the second system is schematically presented in Fig. 8.5. The aqueduct outside the city limits (outer aqueduct) delivered water into a cistern (Fig. 8.6). From this cistern, a canalization made of pipes (roughly oriented parallel to the Cardo Maximus) and an inner aqueduct (inside the city limits, running towards the Cardo Maximus, Fig. 8.7) started. Just before the entrance in this cistern, a derivation was realized. This derivation brought water into another cistern (cistern 2 in Fig. 8.5), located inside a guard tower. The date of construction of this other cistern is not precisely known, but it was built at least in the beginning of the 5th century.

Fig. 8.4 Excavated terracotta pipes used during phase 1. Color version available in Appendix

8.1.2.3 Phase 3

The third and most recent hydraulic system was built at the beginning of the 6th century A.D. and was used until the 7th century A.D. (Byzantine Period) (Viviers and Vokaer, 2009).

In Fig. 8.8, the flow of water in this third hydraulic system is presented. The main elements of this system are numbered in Fig. 8.8. In this system, the roles of the two cisterns appearing in the second hydraulic system are permuted (and so are their numbers in Figs. 8.5 and 8.8).

The outer aqueduct (point 1 in Fig. 8.8) is blocked at point 2 and the water was derived into the large cistern in the guard tower (Figs. 8.9 and 8.10). From this cistern, the water was carried by an inner aqueduct (point 3 in Fig. 8.8; Figs. 8.9 and 8.11), roughly oriented North-South, parallel to the Cardo Maximus, along the second East Street. From point 4, the covering blocks are missing. An observation hole is observed at point 5. From this inner aqueduct, several derivations, or branches, that were delevering water elsewhere in the city, are observed.

A first derivation is observed approximately 10 m after the beginning of the aqueduct (Fig. 8.9). It is mainly composed of a canalization made of the fitting of basic pipes into each other. Just after the beginning of the derivation, a room of visit is observed (point 6 in Fig. 8.8; Fig. 8.12). A water repellent coating made of mortar is observed on the room walls. The connection between the aqueduct and the room of visit is dug through the aqueduct wall. It is 34 cm long and has a rectangular section of 19×10 cm^2. The first pipe of the canalization is connected to this room. This derivation was carrying water into the cistern where the water was delivered by the outer aqueduct during phase 2 (Figs. 8.9 and 8.13). It is thought that the water in this cistern was used for the work of craftsmen. A 90° bend is observed at point 7. It is realized by the connection of pipes on a hollow stone. This first derivation has been built in the 6th century (Viviers and Vokaer, 2009).

A second derivation is observed approximately 5 m after the first one. The end of this second derivation has not been excavated yet, but it seems to end near or even beyond the Cardo Maximus, 80 m from the aqueduct. It is also mainly composed of

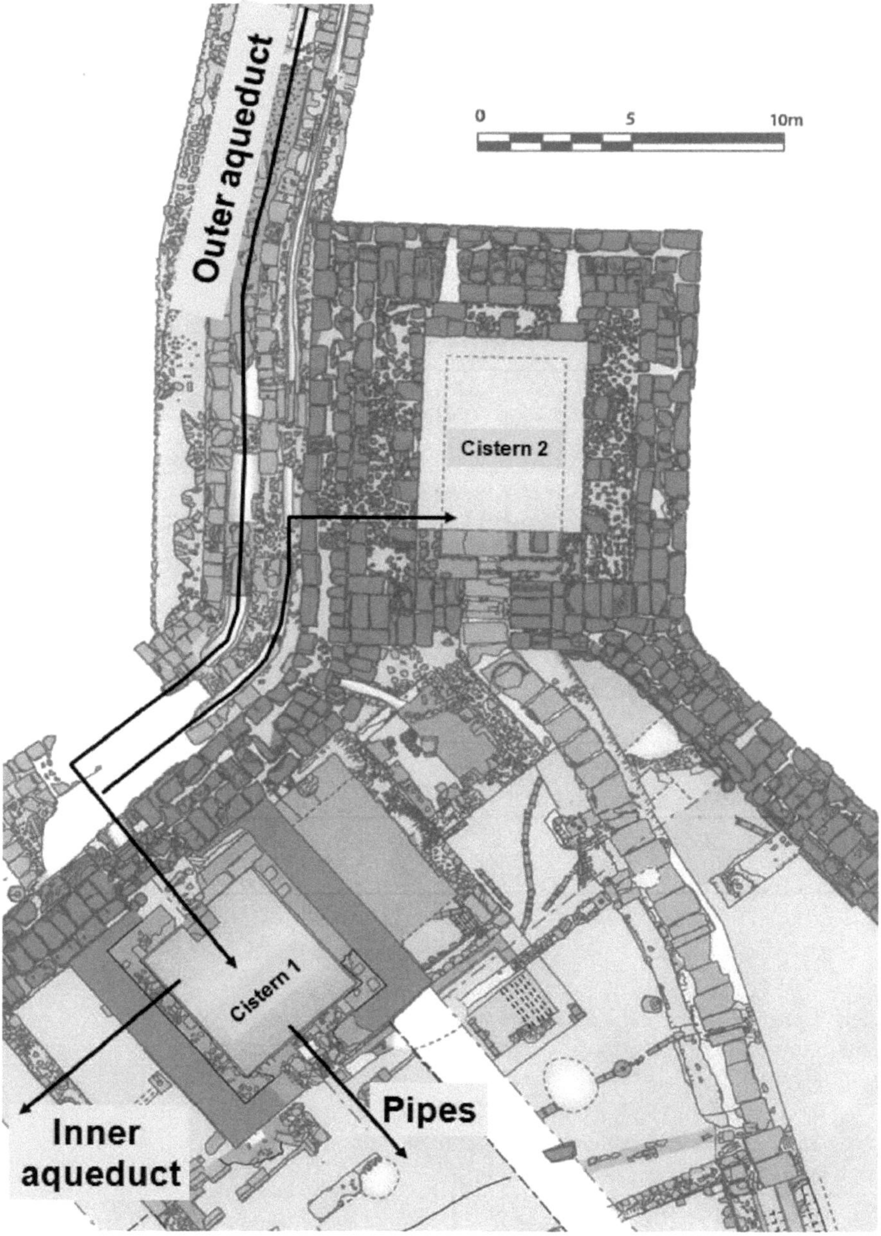

Fig. 8.5 Flow of the water in the second water system

a canalization made of the fitting of basic pipes into each other. Just after the beginning of the derivation, a room of visit is observed (point 8 in Fig. 8.8; Fig. 8.14). The connection between the aqueduct and the room of visit is made of two lead pipes crossing the aqueduct wall, each one having a rectangular section of $6 \times 4\ \text{cm}^2$.

Fig. 8.6 Outer aqueduct delivering water into a cistern. Color version available in Appendix

Fig. 8.7 Aqueduct inside the city limits. Color version available in Appendix

The first pipe of the canalization is connected to this room. A few meters after the beginning of the derivation, a terracotta decanter was excavated (point 9 in Fig. 8.8; Fig. 8.15).

These excavations show that the Byzantine city was not only using the Roman aqueduct but was also able to rebuild a new water supply system. The efficiency of this system is analyzed in this chapter.

8.1.3 Objectives of this Work

The Romans had a remarkable engineering knowledge of water supply (Viollet, 2000). Water was carried to the Roman cities through aqueducts that could reach more than 100 km. Within the cities, the water was distributed through a complex

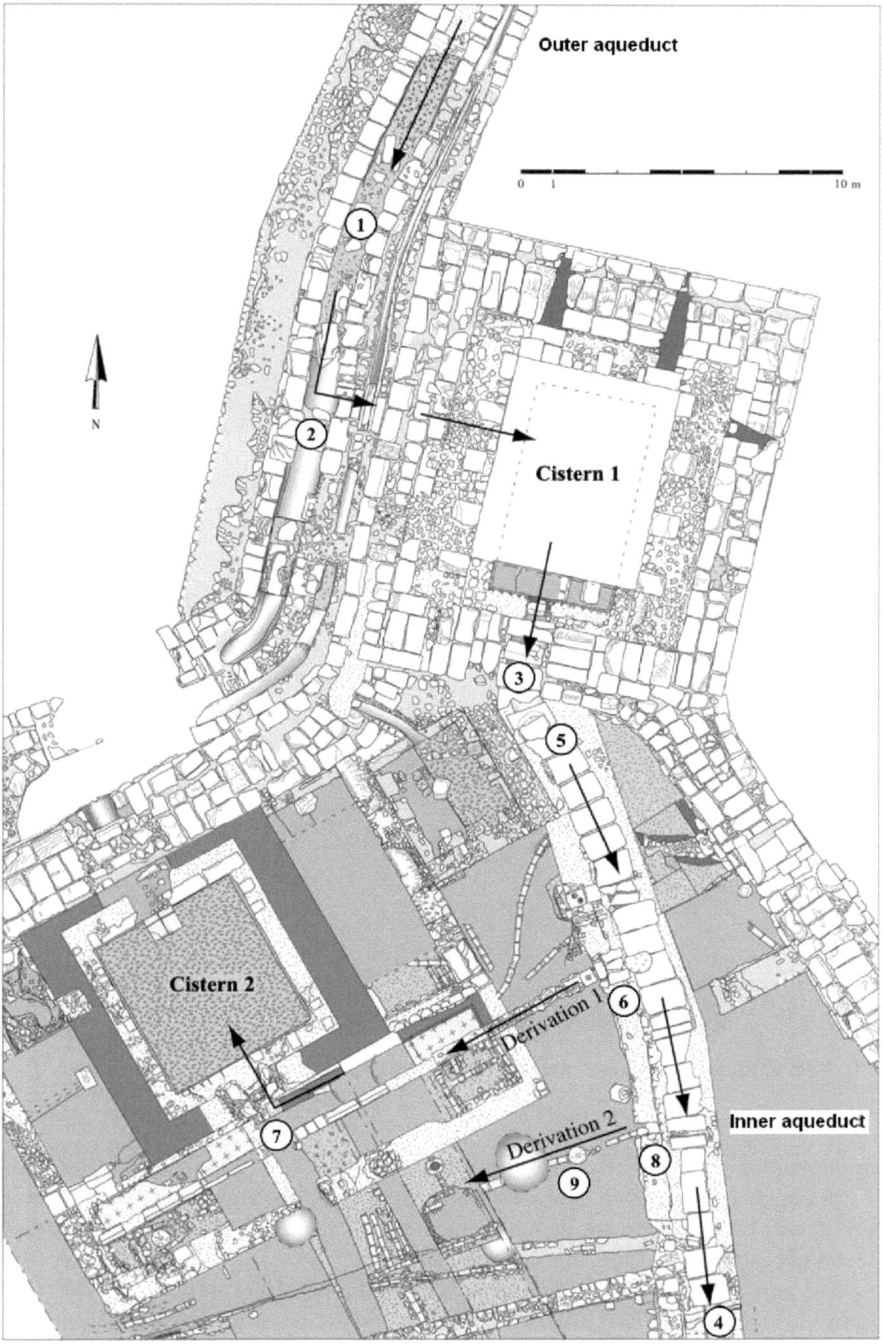

Fig. 8.8 Flow of water in the most recent water system

Fig. 8.9 Outer aqueduct, cistern inside the guard tower, inner aqueduct, first derivation and cistern at the end of the first derivation. Color version available in Appendix

Fig. 8.10 Entrance of the water in the guard tower (point 2 in Fig. 8)

combination of water towers and pipes. Wastewater was evacuated from the cities by drainage systems. Only a few Roman writings on this engineering practice were preserved, but archaeology, as in Apamea, offers some precise illustration of their techniques.

The surviving written records of Frontinus (Evans, 1994) (Herschel, 1973) and Vitruvius (Morgan, 1960) provides some understanding of water supply systems in Roman times. While these works give insight into the design methodology of water supply systems of that period, they also reflect pre-scientific views of hydraulic

Fig. 8.11 Picture taken from the inside of the inner aqueduct. Calcareous deposits on the aqueduct walls can be observed

Fig. 8.12 Room of visit in the first derivation. The aqueduct is on the *left* of this picture

principles (Ortloff and Crouch, 2001). For instance, in the work of Frontinus, a concept such as the flow rate is not known.

As a modern resource, Computational Fluid Mechanics software provides researchers and archeologists the tools to unearth the engineering details of the Roman water supply systems. These computational means allow the simulation of the flow of water through archaeological remains that are well preserved, opening

Fig. 8.13 End of the first derivation in a cistern. Color version available in Appendix

Fig. 8.14 Remains of the room of visit in the second derivation. The aqueduct is at the rear of this picture

new possibilities for the analysis of these ancient water systems. An analysis of this type has been performed for Apamea and is presented in this chapter. The political and administrative status of Apamea has allowed archaeologists to piece together an excellent picture of the most advanced Roman technology in late antiquity.

Fig. 8.15 Terracotta decanter excavated in the second derivation

The *first objective* of the work presented in this chapter is to characterize the operation of the third hydraulic system (Byzantine period). The water flow rate in the inner aqueduct is computed using the Manning equation. The flow rate and the energy losses in each of the two derivations are computed using global energy balances, coupled with local CFD simulations, realized with the commercial software Fluent 6.3. The time needed to fill the cistern is also evaluated. The results of this characterization, presented in Section 8.4, provide a new quantitative description of the water supply system of the city, supplementing the usual field observations. The methodology followed for this characterization is presented briefly in Section 8.2 and deeply in Appendix 1, while the CFD simulations realized are presented in Section 8.3.

The *second objective* of this work is to analyse the results of the hydraulic characterization from an archaeological point of view. Three open historical questions about Apamea are tackled. The first one is whether or not the aqueduct was the only water supply system of the town. The second question is whether or not the Byzantine city was able to develop new efficient water systems. The third question is whether or not, during the Byzantine period, the hydraulic system was able to deliver water beyond the Cardo Maximus. Today, none of these questions has a clear answer (Balty, 2000) (Viviers and Vokaer, 2009).

It is shown in Section 8.5 that the analysis of the data provided by the hydraulic characterization provides new interesting elements regarding the answer to these questions. It demonstrates the ability of this modern approach, coupled with classical field observations, to analyse ancient well preserved remains.

8.2 Methodology

The methodology followed for the characterization of the operation of the third hydraulic system is briefly presented here. A complete description of this methodology is presented in Appendix 1.

8.2.1 Flow in the Inner Aqueduct

The flow rate in the inner aqueduct is written Q_{aq} (kg/s). It can be calculated using the Manning equation (Whitaker, 1968):

$$\frac{Q_{\mathrm{aq}}}{\rho bh} = V_{\mathrm{aq}} = \left(\frac{bh}{b+2h}\right)^{2/3} \sqrt{\frac{2\,g\tan(\theta)}{29\,n^2}} \tag{8.1}$$

where h is the water height in the inner aqueduct, b is the aqueduct width, ρ is the volumetric mass of water, g is the acceleration of the gravity and $\tan(\theta)$ is the aqueduct slope. n is the Manning coefficient, depending on the nature of the aqueduct walls.

b and $\tan(\theta)$ can be measured. h can also be measured as a calcareous deposit is observed in the aqueduct. Values of n are tabulated (Bailhache, 1979; Lencastre, 1995; Whitaker, 1968).

8.2.2 Flow in the Derivations

In Fig. 8.16, the first derivation is schematically presented. Three important points of the flow in the derivation are presented. Point A is located in the connection between the aqueduct and the room of visit. Point B is located in the first pipe of the canalization, connected to the room of visit. Point C is located in the last pipe of the canalization, just before the entrance in the cistern.

As presented in Appendix 1, the expression of the flow energy difference between points A and B and points B and C allows to determine an equation for the flow rate in the derivation and therefore for the time evolution of the water height in the cistern at the end of the derivation:

$$Q = \rho\Omega_{\mathrm{cit}}\frac{dH}{dt}\begin{cases} \rho\Omega_{\mathrm{B}}\sqrt{2\,g\dfrac{h+D_1}{1+\frac{1}{2}\left(\frac{\Omega_{\mathrm{B}}}{\Omega_{\mathrm{A}}}\right)^2+L_1C_{\mathrm{C1}}+C_{\mathrm{B}}+C_{\mathrm{R1}}\left(\frac{\Omega_{\mathrm{B}}}{\Omega_{\mathrm{A}}}\right)^2}}, & \text{when } H>H_{\mathrm{d}} \\ \rho\Omega_{\mathrm{B}}\sqrt{2\,g\dfrac{h+D_1-(H-H_d)}{1+\frac{1}{2}\left(\frac{\Omega_{\mathrm{B}}}{\Omega_{\mathrm{A}}}\right)^2+L_1C_{\mathrm{C1}}+C_{\mathrm{B}}+C_{\mathrm{R1}}\left(\frac{\Omega_{\mathrm{B}}}{\Omega_{\mathrm{A}}}\right)^2}}, & \text{when } H>H_{\mathrm{d}} \end{cases} \tag{8.2}$$

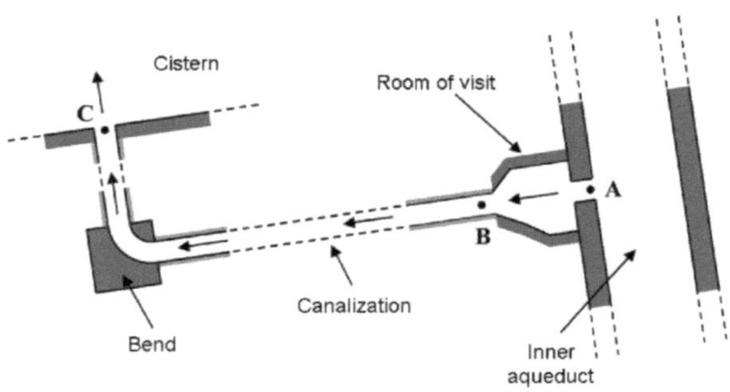

Fig. 8.16 Schematic representation of the first derivation

where Q is the flow rate in the derivation, Ω_{cit} is the ground surface of the cistern, H is the height of water in the cistern, H_{d} is the height difference between the end of the canalization and the bottom of the cistern (see Fig. 8.13), D_{l} is the height difference between point C and point A (point A is higher than point C), Ω_{A} is the area of the section of the connection between the room of visit and the aqueduct and Ω_{B} is the area of the section of the pipes in the canalization, C_{Cl} is the energy drop coefficient of the flow in 1 m of the canalization, L_{l} is the length of the canalization, C_{B} is the energy drop coefficient of the flow in the bend, C_{Rl} is the energy drop coefficient of the flow between points A and B.

L_{l}, Ω_{cit}, h, H_{d}, D_{l}, Ω_{B}, Ω_{A} can be measured. Therefore, if C_{Cl}, C_{Rl} and C_{B} can be determined, Equation (8.2) allows to calculate the water flow rate in the derivation and to predict the time evolution of the water level in the cistern.

In this work, the energy drop coefficients are calculated using either semi-empirical correlations or a Computational Fluid Dynamics commercial software: Fluent 6.3. In this software, the local equations of fluid mechanics (i.e. the continuity equation and the Navier-Stokes equations) can be solved numerically in complex geometries. From the solution of these equations, energy losses, and hence energy drop coefficients, can be calculated. A short description of the modelling and the simulation of turbulent flows is provided in Appendix 1.

A similar methodology is followed for the characterization of the flow in the second derivation.

8.3 Energy Losses in the Different Elements of the Derivations

8.3.1 Rooms of Visit

8.3.1.1 Derivation 1

The geometry of the room of visit in the first derivation is presented in Fig. 8.17.

Flow rates Q between 2 and 25 kg/s are studied. A 3D mesh of approximately 170 10^3 hexahedral cells is generated.

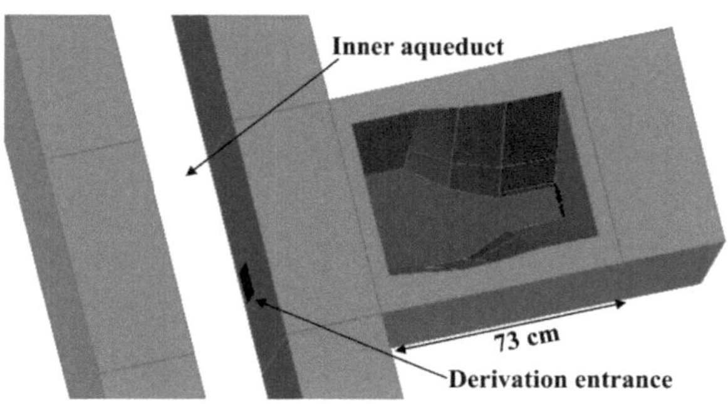

Fig. 8.17 Geometry of the room of visit in the first derivation, drawn with Gambit

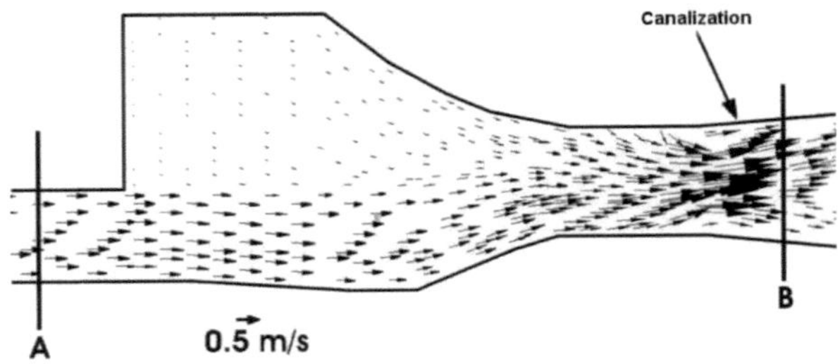

Fig. 8.18 Velocity vector field in a horizontal plane 4 cm above the bottom of the room of visit in the first derivation. Q = 10 kg/s

The level of water in the room of visit is roughly determined from the measurement of the height of a calcareous deposit (+/− 30 cm). An accurate determination of this level is not essential, as the energy loss in the room of visit is not very sensitive to it.

The standard k-ϵ turbulence model is first used. Once the convergence is reached, the turbulence model is switched to a k-ϵ RNG model. Further iterations are performed, until convergence is reached again.

In Fig. 8.18, the simulated velocity vector field in a horizontal plane in the room of visit (4 cm from the bottom) is presented, for $Q = 10$ kg/s. The presence of a large stagnant zone can be observed.

When a simulation has been realized, the volumetric energy difference between cross-sections A and B, defined in Fig. 8.18, is reported. In Fig. 8.19, $E_{\mathrm{A}} - E_{\mathrm{B}}$ is plotted as a function of Q^2. A linear relationship between the energy loss and Q^2 is observed. A linear regression of these data gives $C_{\mathrm{R1}} = 6.9$ (using Equation (8.9) and the fact that Ω_{A} is measured equal to 190 cm^2).

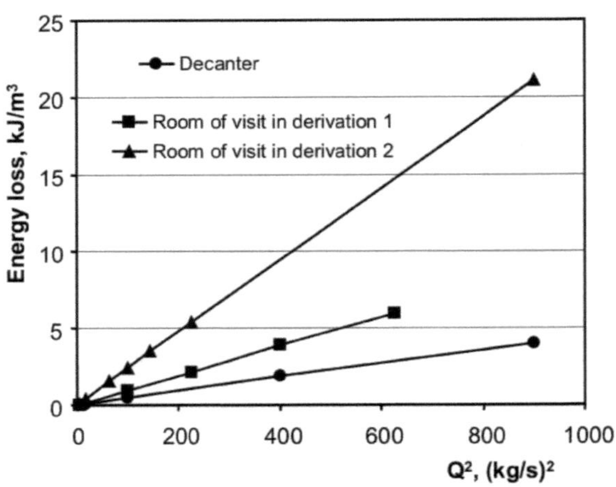

Fig. 8.19 Energy loss in rooms of visit, as functions of Q^2. Energy loss in the decanter in the second derivation, as a function of Q^2

8.3.1.2 Derivation 2

Flow rates Q between 2 and 30 kg/s are studied. The flow in the room of visit is turbulent. A 3D mesh of approximately 170×10^3 hexahedral cells is generated. A k-ϵ standard model of turbulence is used here.

In Fig. 8.20, the structure of the flow in a horizontal plane in the room of visit (12 cm from the bottom) is presented, for Q = 8 kg/s. The presence of a large vortex can be observed.

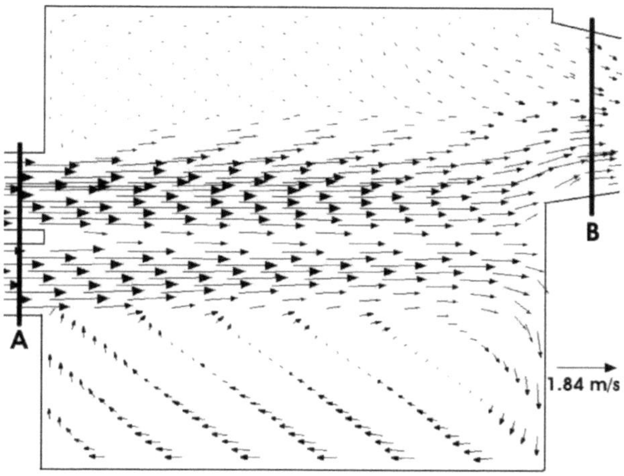

Fig. 8.20 Velocity vector field in a horizontal plane 12 cm above the bottom of the room of visit in the second derivation. Q = 8 kg/s

When a simulation has been realized, the volumetric energy difference between cross-sections A and B, defined in Fig. 8.20, is reported. In Fig. 8.19, this energy loss is plotted as a function of Q^2. A linear relationship between this energy loss and Q^2 can be observed. The energy drop coefficient of the flow in the room of visit is written C_{R2}. $C_{R2} = 1.1$ can be extracted from the data of Fig. 8.19 (using the fact that the area of the section of the connection between the room of visit and the aqueduct is equal to 48 cm^2).

It can be observed that, at a same Q, the energy loss in the room of visit in the second derivation is more than twice the energy loss in the room of visit in the first one (Fig. 8.19). A part of this significant difference is due to the presence of the large vortex in the room of visit in the second derivation.

8.3.2 *Canalizations*

The canalizations of the two derivations are composed of the fitting of basic pipes into each other. Therefore, they are characterized by the repetition of an elementary pattern. These elementary patterns are presented in Figs. 8.21 and 8.22. The flow in the canalizations is entirely described by the flow in these elementary patterns. The basic pipes are made of clay and the connections between them are made of mortar. The clay in contact with the water has been glazed. Corresponding roughness is

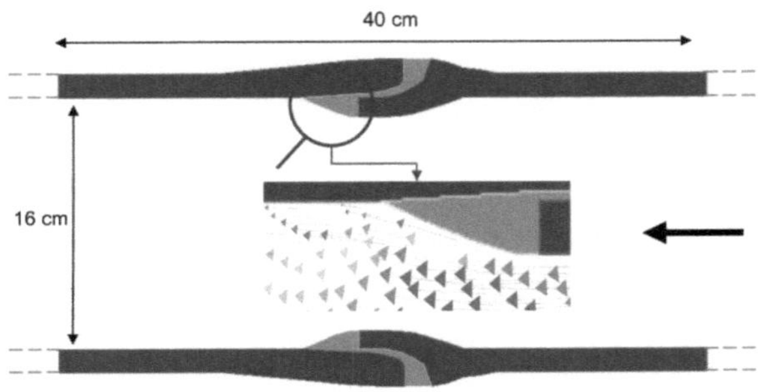

Fig. 8.21 Elementary pattern of the canalization of the first derivation (the mortar is in *light grey*). Velocity vector field near the connection between the pipes of the derivation. Q = 10 kg/s, K = 0

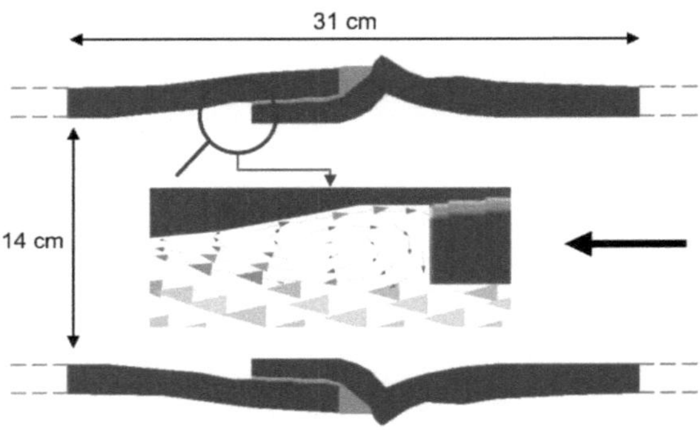

Fig. 8.22 Elementary pattern of the canalization of the second derivation (the mortar is in *light grey*). Velocity vector field near the connection between the pipes of the derivation. Q = 10 kg/s, K = 0

between 10 and 30 μm. A calcareous deposit is produced by the flow of water in the canalizations. Its roughness is between 0.2 and 0.5 mm.

Flow rates Q between 2 and 30 kg/s and roughness K between 0 and 0.5 mm are studied. The flow in the canalizations is turbulent, as their Reynolds numbers are larger than 10^4.

The elementary patterns are axisymmetric. Hence, the simulation of the flow in these geometries can be reduced to a 2D problem.

The k-ϵ RNG turbulence model is used. Several other models have been tested. Besides the k-ϵ standard model, they give all similar results. Periodic boundary conditions for the velocity vector are imposed at the inlet and the outlet of the elementary patterns.

In Figs. 8.21 and 8.22, the velocity vector field near the connection between the basic pipes is presented for the two derivations. In the second derivation, the presence of a vortex generated by this connection can be observed. In the first derivation, the shape given to the mortar prevents the appearance of this vortex.

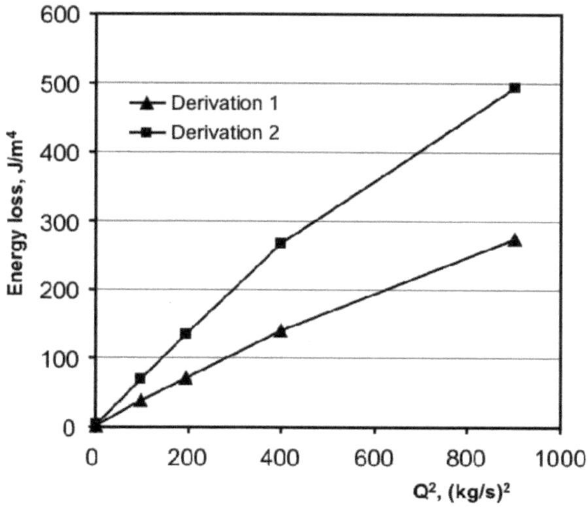

Fig. 8.23 Energy loss in the canalizations of the two derivations, as functions of Q^2. $K = 0$

When a simulation has been realized, the energy loss per unit length of the canalization can be easily calculated. In Fig. 8.23, this energy loss is plotted as a function of Q^2 for the two canalizations and for $K = 0$. Linear relationships between the energy losses and Q^2 can be observed. The energy drop coefficient of the flow in 1 m of the canalization in the second derivation is written C_{C2}. $C_{C1} = 0.26\ \text{m}^{-1}$ and $C_{C2} = 0.27\ \text{m}^{-1}$ can be extracted from the data of Fig. 8.23 (using the fact that the area of the section of the pipes in the first and the second derivation are measured equal to 201 and 154 cm^2, respectively).

At a same Q, the energy loss in the canalization of the second derivation is almost twice the energy loss in the canalization of the first one (per unit length of the canalization). The energy loss associated with the vortex located at the connection between the basic pipes of the second derivation is mainly responsible for this difference.

In Fig. 8.24, the energy loss per unit length of the canalization is plotted as a function of the roughness, for the two derivations and for $Q = 20$ kg/s. For the roughness of the glazing, the pipes are hydrodynamically smooth. When a calcareous deposit appears in the canalizations, the energy loss is significantly modified. When $K = 350\ \mu\text{m}$, it is increased by 37% in the first derivation and by 14% in the second derivation.

8.3.3 Bend in Derivation 1

The 90° bend met at point 7 (Fig. 8.8) is realized by the connection of pipes on a hollow stone. The way the stone is dug is presented in Fig. 8.25. The cross-section of the flow remains almost circular through the entire bend. Its area remains constant.

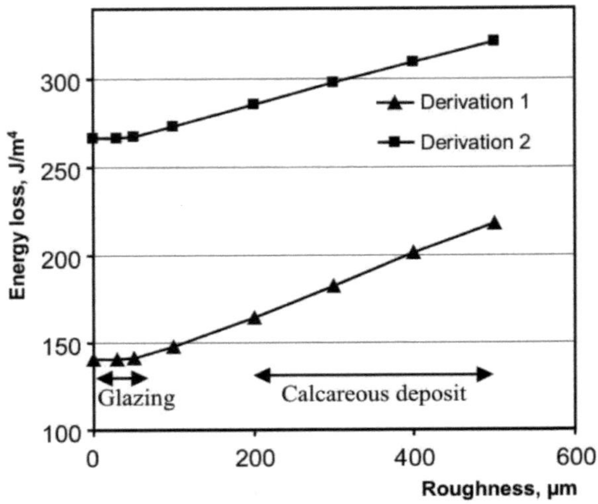

Fig. 8.24 Energy loss in the canalizations of the two derivations, as functions of K. Q = 20 kg/s

Fig. 8.25 Geometry of the 90° bends in the first derivation

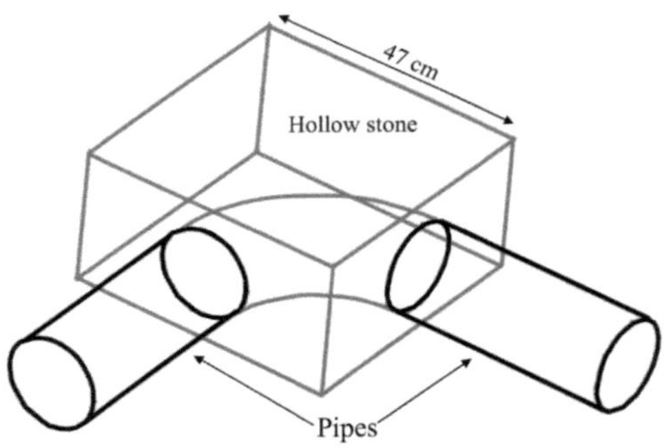

The energy loss associated with this bend can be calculated using experimental data (Lencastre, 1995). It appears that this energy loss is small compared with energy losses in other parts of the derivation. It is close to the energy loss in 1 m of the canalization. Therefore, C_B is neglected in front of C_{R1} and $L_1 C_{C1}$ in Equation (8.2).

8.3.4 Decanter in Derivation 2

The geometry of the decanter in the second derivation is presented in Fig. 8.26. A water repellent coating made of mortar is found on the walls of the decanter.

Flow rates Q between 4 and 30 kg/s are studied. The flow in the decanter is turbulent. A 3D mesh of approximately 120×10^3 hexahedral cells is generated. A turbulence model k-ϵ Realizable is used here.

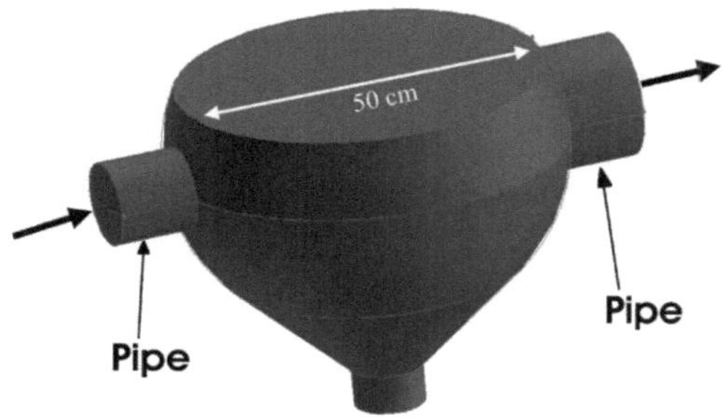

Fig. 8.26 Geometry of the decanter excavated in the second derivation, drawn with Gambit

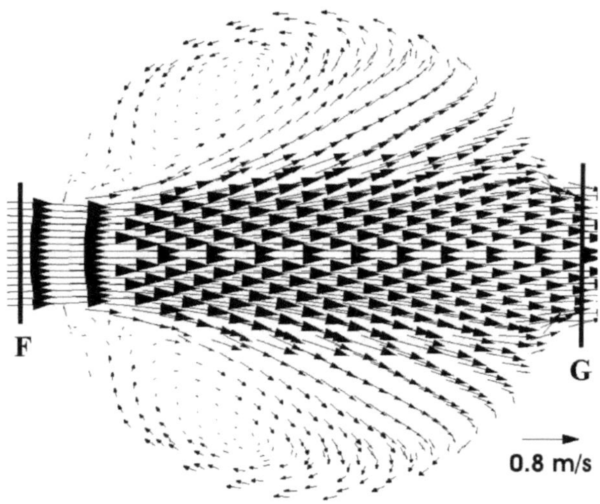

Fig. 8.27 Velocity vector field in a horizontal plane in the decanter. Q = 10 kg/s

In Fig. 8.27, the velocity vector field in a horizontal plane in the decanter is presented.

When a simulation has been realized, the energy difference between cross-sections F and G, defined in Fig. 8.27, is reported. In Fig. 8.19, this energy loss is plotted as a function of Q^2. A linear relationship between this energy loss and Q^2 can be observed. The energy drop coefficient of the flow in the decanter is written C_D. $C_D = 2.1$ can be extracted from the data in Fig. 8.19.

8.4 Numerical Description of the Flow in the Third Hydraulic System

The first objective of the work presented in this chapter is to characterize, using the methodology presented in Section 8.2 and the simulations presented in Section 8.3, the operation of the third hydraulic system. This characterization is presented in this section.

8.4.1 Inner Aqueduct

Between the observation hole (point 5 on Fig. 8.8) and the end of the covered part of the aqueduct (point 4 on Fig. 8.8), a slope of 2.1 ± 0.3 mm/m is measured. Six measurements of the aqueduct width are performed between the entrance of the aqueduct and a point located 2 m beyond the observation hole. $b = 57 \pm 2$ cm is obtained.

A calcareous deposit can be observed on the aqueduct walls. Its thickness seems uniform on the lateral walls. Hence, h, the water height in the canal, had slow variation in time. The average height reached by the deposit on the aqueduct lateral walls is measured four times and $h = 92 \pm 1$ cm is obtained. It shows that the water level was a few centimetres below the covering blocks.

For such a calcareous deposit, $n = 0.014\ \mathrm{m}^{1/6}$ can be taken (Bailhache, 1979; Lencastre, 1995; Whitaker, 1968). The application of Equation (8.1) yields $Q_{\mathrm{aq}} = 512 \pm 61$ kg/s.

At the beginning of its use, the aqueduct carried a same mass flow rate, but the roughness of the walls was smaller, as there was no calcareous deposit yet. In this case, the water is in contact with a water repellent coating made of mortar, for which $n = 0.01\ \mathrm{m}^{1/6}$ can be taken (Bailhache, 1979; Lencastre, 1995; Whitaker, 1968). Setting $Q_{\mathrm{aq}} = 512$ kg/s and $n = 0.01\ \mathrm{m}^{1/6}$ in Equation (8.1) (where θ and b keep their previous values) yields $h = 66$ cm.

8.4.2 First Derivation

Equation (8.2) allows to calculate the water flow rate in the derivation and to predict the time evolution of the water level in the cistern at its end if L_1, h, D_1, Ω_{cit}, Ω_{A}, Ω_{B}, H_{d}, C_{C1}, C_{B} and C_{R1} are known.

$L_1 = 12$ m, $h = 92$ cm, $D_1 = 4$ cm, $\Omega_{\mathrm{cit}} = 24.75\ \mathrm{m}^2$, $\Omega_{\mathrm{A}} = 190\ \mathrm{cm}^2$, $\Omega_{\mathrm{B}} = 201\ \mathrm{cm}^2$ and $H_{\mathrm{d}} = 1.7$ m, have been measured and C_{R1}, C_{B} and C_{C1} have been calculated previously.

When the level of water in the cistern is below the end of the canalization ($H < H_d$), application of Equation (8.2) yields $Q = 25$ kg/s, when the pipes are hydrodynamically smooth (no calcareous deposit) and $Q = 23$ kg/s when a calcareous deposit with a 350 μm roughness is present on the canalization walls.

The values of Q with and without the calcareous deposit are very close. As C_{C1} is significantly modified by the presence of the calcareous deposit (see Fig. 8.23), it indicates that a large part of the energy loss is not located in the canalization. When there is no calcareous deposit, the energy loss associated with the entrance in the derivation is 0.4 kJ/m^3, the energy loss in the room of visit is 5.9 kJ/m^3 and the energy loss in the canalization is 2.4 kJ/m^3.

Q_{max} is defined as the flow rate that would be obtained if the derivation could transport the water without any energy loss. Setting all energy losses to zero in Equation (8.2) yields an equation for Q_{max}. $Q_{max} = 87$ kg/s is calculated. Therefore, the efficiency Q/Q_{max} of the derivation is close to 0.3, when there is no calcareous deposit on the canalization wall.

Equation (8.2) allows calculating that approximately 30 minutes are needed to fill the cistern up to the end of the derivation.

When the level of water in the cistern is above the end of the canalization ($H > H_d$), the flow rate in the derivation gradually decreases. When $Q = 0$, the maximum value of H, written H_{max}, is reached. Setting $Q = 0$ in Equation (8.2) yields $H_{max} = 2.66$ m. The cistern can be filled with 66 m^3 of water. It is calculated that it was taking approximately an hour to fill the cistern completely.

8.4.3 Second Derivation

The end of the second derivation has not been excavated yet. Therefore, a hypothetic derivation, but coherent with the excavated part of the second derivation, is considered here. This hypothetic derivation ends close to the Cardo Maximus (80 m from the aqueduct) and its canalization runs on ground level straightforwardly to the Cardo Maximus. Once there, the water can be brought back to the first floor of a building, or even to a building on the other side of the street, with a few supplementary metres of canalization and with some 90° bends (with no energy loss associated). The end of the derivation was not submerged. Therefore, an equation similar to Equation (8.2) can be used to calculate the flow rate delivered by the derivation, except that the energy loss associated with the decanter has to be taken into account in the energy balance.

In Fig. 8.28, the flow rate Q in the hypothetic derivation is plotted as a function of D_2, the height difference between the beginning of the derivation and its end. D_2

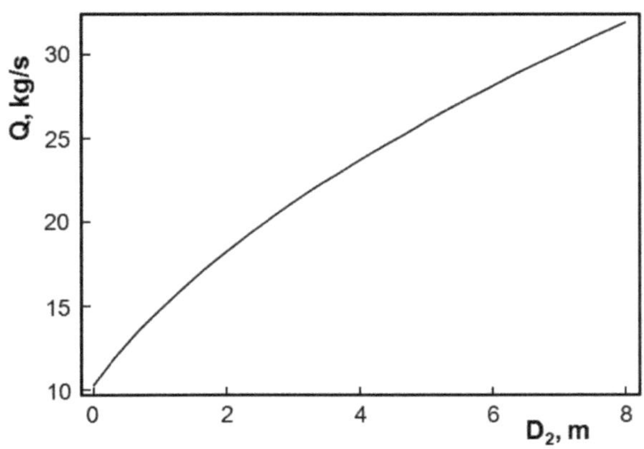

Fig. 8.28 Mass flow rate in the second derivation as a function of D_2

equals approximately 8 m if the canalization ends on the ground level, while D_2 equals approximately 4 m if the canalization ends on the first floor of a building.

When $D_2 = 4$ m, $Q = 24$ kg/s is obtained, when the pipes are hydrodynamically smooth. The efficiency Q/Q_{max} of the derivation is equal to 0.16. The energy loss associated to the entrance in the derivation is 6.1 kJ/m^3, the energy loss in the room of visit is 13.2 kJ/m^3, the energy loss in the decanter is 2.5 kJ/m^3 and the energy loss in the canalization is 25.2 kJ/m^3.

8.5 Analysis of the Flow in the Third Hydraulic System

The second objective of the work presented in this chapter is to analyse the results of the hydraulic characterization from an archaeological point of view. This analysis is presented in this section. The focus is given on three open historical questions about Apamea. The first one is whether or not the aqueduct was the only water supply system of the town. The second question is whether or not the Byzantine city was able to develop new efficient water systems. The third question is whether or not, during the Byzantine period, the hydraulic system was able to deliver water beyond the Cardo Maximus.

8.5.1 Inner Aqueduct

It has been shown that the aqueduct carried a water flow rate of 0.5 m^3/s or 43.2×10^3 m^3/day. Nowadays, there is an open question about the water consumption in Roman cities. Daily consumptions between 200 and 500 l per capita are reported (Chanson, 2002; Viviers and Vokaer, 2009). According to these values, 43×10^3 m^3 of water per day could fulfilled the needs of a population between 90 and 220 thousand people.

It is estimated that, during the 6th century, the number of inhabitants of Apamea was between 100 and 200 thousand people (Viviers and Vokaer, 2009). Therefore, the evaluated flow rate delivered by the aqueduct is a new element that leads to think that the aqueduct was the only water supply system of the town.

This new element strengthens the affirmation, made by some authors (Viviers and Vokaer, 2009), that the site topography would have made difficult the building of another water adduction, as it should have been weaker because of the necessity of crossing the wadi or the Ghab valley.

At the beginning of the use of the aqueduct, the water was in contact with a smooth water repellent coating made of mortar. The water level in the aqueduct was 66 cm height. As soon as a calcareous deposit was formed, the roughness of the surface increased and hence, the level of water in the aqueduct rose. It finally stabilized at 92 cm, only a few centimetres below the covering blocks. This observation could lead us to think that the Romans had foreseen this elevation of water height at least qualitatively and maybe even quantitatively.

8.5.2 First Derivation

This first derivation carried a water flow rate of 25 kg/s towards a cistern that took approximately one hour to be filled with 66 m^3 of water.

Regarding the transport of water, each part of the derivation has good characteristics.

The shape given to the mortar, ensuring the fitting of the basic pipes of the canalization into each other, prevents the appearance of vortices (see Figs. 8.21 and 8.22), and hence decreases the energy loss in the canalizations. The glazing of the pipes ensures the walls to be hydrodynamically smooth walls (at least in the absence of calcareous deposit). The energy drop coefficient C_{C1} (Equation (8.10)) is equal to 0.26 m^{-1} when there is no calcareous deposit.

It can be pointed out once again that the energy loss associated with the 90° bend is negligible in comparison with energy losses in other parts of the derivation. Thus, it has almost no influence on the flow rate in the derivation. The engineering technique of this kind of bends had not to be improved.

No vortex is generated in the room of visit. The flow is conducted from the entrance to the exit by the room walls. However, a stagnant zone is observed in the room (see Fig. 8.18). Such a stagnant zone consumes a lot of energy and indeed, the major part of the energy loss in the derivation is located in this room.

Finally, the connection between the aqueduct and the room of visit has a large cross-section. Hence, the energy loss associated with this entrance is negligible (see Equation (8.3)), compared to energy losses in other parts of the derivation.

Q/Q_{max} is close to 0.3. Therefore, in the total absence of energy loss in the entire derivation (physically impossible as energy is needed to sustain the turbulent flow of the water) the cistern would take approximately 20 minutes to be filled instead of 1 hour.

It can be computed that a similar derivation built with modern techniques and materials would have an efficiency of 0.72. To calculate this value, several assumptions have been made. It is first assumed that the modern derivation does not include a room of visit, the canalization is directly connected to the aqueduct. It is further assumed that the canalization is made of plastic pipes having a diameter of 16 cm. It is also assumed that no energy loss occurs at the connection between the pipes. Finally it is assumed that the 90° bend is constructed following actual rules. The data and correlations needed to compute the efficiency of this modern derivation have been found in (Lencastre 1995).

In this modern derivation, the energy drop coefficient C_{C1} of the flow in 1 m of the canalization is equal to 0.07 m^{-1}. It has been calculated using the Colebrook-White formula (Lencastre, 1995). At a same flow rate, the energy loss in the modern derivation would be three times smaller than this energy loss in the roman derivation.

It can finally be pointed out that the position of the cistern entrance allows a natural regulation of the incoming flow rate. As long as the level of water in the cistern is below the end of the canalization, the incoming flow rate is constant in time. When the level of water in the cistern is above the end of the canalization, the incoming flow rate is continuously decreased in time until a zero flow rate is reached.

This analysis clearly demonstrates that the first derivation, built in the 6th century, is a very good work, reflecting an excellent technical knowledge of water supply, until the end of antiquity. The Byzantine city was not only using the Roman aqueduct but was also able to rebuild a new efficient water supply system. Therefore, the analysis of the efficiency of this first derivation provides a new interesting element regarding the answer to the second open historical question about Apamea tackled in this chapter.

8.5.3 Second Derivation

The relation between the flow rate Q and the final height of a hypothetic second derivation, ending close to the Cardo Maximus, has been established (see Fig. 8.28). It is shown that if the derivation ends on the first floor of a building close to the Cardo Maximus, the flow rate in the derivation would be approximately 24 kg/s.

Therefore, the analysis of the second derivation demonstrates that, during the 6th century, it was technically possible to provide large amount of water to the part of the city located beyond the Cardo Maximus. This gives a new interesting element for the answer to the third historical questions about Apamea tackled in this chapter.

Regarding the transport of water, the technical design of this derivation is clearly not as good as the one of the first derivation. This observation is valid for each part of the derivation.

The energy loss in the room of visit in the second derivation is, at a same flow rate, more than twice the energy loss in the room of visit in the first derivation (Fig. 8.19). There is no shaping of the room walls, leading to the formation of a large vortex that consumes a lot of energy (Fig. 8.20).

At a same flow rate, the energy loss per unit length of the canalization in the second derivation is almost twice the energy loss per unit length of the canalization in the first derivation (Fig. 8.24). The vortex generated at the connection between basic pipes is mainly responsible for this difference (Fig. 8.22). In the first derivation, it is the shape given to the mortar that prevents the appearance of this vortex (Fig. 8.21).

At a same flow rate, the energy lost at the entrance of the second derivation is 15 times larger than the energy lost at the entrance of the first derivation. In the second derivation, the area of the section of the connection between the room of visit and the aqueduct is equal to 0.0048 m^2, while in the first derivation it is equal to 0.019 m^2. As it can be seen in Equation (8.3), the energy lost at the entrance of a derivation increases with an increasing velocity in the connection between the room of visit and the aqueduct, and therefore with a decreasing area of the section of this connection. The energy loss associated to the entrance in the second derivation is more than 10 percent of the total energy loss in the derivation.

Some words can be said about the decanter. The first question arising concerns the utility of this decanter. A possible answer is that it could have been placed there to eliminate solid particles produced by the erosion of the walls in the room of visit. Indeed, no water repellent coating has been found in this room. It could have been omitted in the reconstruction of this room later, after its destruction by an earthquake.

The second question arising concerns the efficiency of this decanter. As it can be seen in Fig. 8.27, the main part of the flow shortcuts the decanter, with a residence time of approximately 0.6 s. Hence, the decantation of solid particles in suspension is unlikely to take place in the decanter, only larger particles, sedimented in the bottom of the canalization and drifted by the flow, can be collected in the decanter.

When $D_2 = 4$ m, $Q/Q_{\max}$ is equal to 0.16. Therefore, in the total absence of energy loss in the entire derivation the flow rate would be approximately six times larger than the real flow rate.

It can be computed that a similar derivation built with modern techniques and materials would have an efficiency of 0.45. To calculate this value, several assumptions are realized. It is first assumed that the modern derivation does not include a room of visit, the canalization is directly connected to the aqueduct. It is further assumed that the canalization is made of plastic pipes having a diameter of 14 centimetres. It is finally assumed that no energy loss occurs at the connection between the pipes. The data and correlations needed to compute the efficiency of this modern derivation have been found in (Lencastre, 1995).

8.6 Conclusion

The first objective of the work presented in this chapter was to characterize the operation of the hydraulic system operated during the 6th and 7th centuries A.D. (Byzantine Period), composed of the inner aqueduct and the two first derivations.

Several data were computed. The water flow rate in the inner aqueduct was computed using the Manning equation. The flow rate and the energy losses in each of the two derivations were computed using global energy balances, coupled with local Computational Fluid Dynamic (CFD) simulations, realized with the commercial software Fluent 6.3.

Based on these data, the hydraulic behaviour of the studied system was deeply analysed. The two derivations were compared with modern installations.

This hydraulic analysis provides a new description of the water supply system of the city, supplementing the field observations about material, construction, chronology, ...

A second objective of this work was to analyse the results of the hydraulic characterization from an archaeological point of view. Three open historical questions about Apamea were tackled. As presented in the previous section, the results of the hydraulic characterization brought significant new elements regarding the answer to these three questions. The evaluated flow rate delivered by the aqueduct is a new element leading to think that the aqueduct was the only water supply to the town. The analysis of the efficiency of the first derivation clearly demonstrates that the

Byzantine city was able to develop new efficient water supply systems. Finally, the analysis of the second derivation demonstrates that, during the 6th century, it was technically possible to provide water to the part of the city located beyond the Cardo Maximus.

This demonstrates the ability of this modern approach, coupled with classical field observations, to analyse ancient well preserved remains.

Notation

b	inner aqueduct width, m
C_{B}	energy drop coefficient of the flow in the bend in derivation 1, dimensionless
C_{C1}	energy drop coefficient of the flow in 1 m of the canalization in derivation 1, m^{-1}
C_{R1}	energy drop coefficient of the flow between points A and B (see Fig. 8.16), dimensionless
D_1	height difference between point C and point A (see Fig. 8.16), m
E	volumetric energy of the flow, kg/(m s^2)
g	acceleration of the gravity, $\mathrm{m/s}^2$
h	water height in the inner aqueduct, m
H	height of water in the cistern, m
H_{d}	height difference between the end of the canalization and the bottom of the cistern, m
k	turbulent kinetic energy, $\mathrm{m}^2/\mathrm{s}^2$
k_{p}	turbulent kinetic energy at the centre of a cell adjacent to a wall, $\mathrm{m}^2/\mathrm{s}^2$
K	physical roughness of a wall, m
L_1	length of the canalization in derivation 1 (see Fig. 8.16), m
n	Manning coefficient, $\mathrm{m}^{1/6}$
P_{atm}	atmospheric pressure, kg/(m s^2)
P	pressure, kg/(m s^2)
Q	flow rate of water in a derivation, kg/s
Q_{aq}	flow rate of water in the inner aqueduct, kg/s
Q_{max}	maximum flow rate of water that can be delivered by a derivation, kg/s
U_i	component of the velocity along the x_i axis, m/s
U_{p}	velocity magnitude at the centre of a cell adjacent to a wall, m/s
t	time, s
$\underline{V}$	mean velocity in the main direction of the flow, m/s
x_i	ith Cartesian coordinate, m
y_{p}	distance between a wall and the centre of a cell adjacent to this wall, m

Greek Letters

α	see Equation (8.17), s/m
β	see Equation (8.17), m^{-1}
ϵ	dissipation rate of the turbulent kinetic energy, m^2/s^3
ϵ_p	dissipation rate of the turbulent kinetic energy at the centre of a cell adjacent to a wall, m^2/s^3
λ	see Equation (8.19), dimensionless
ν	molecular kinematic viscosity, m^2/s
ν_t	turbulent kinematic viscosity, m^2/s
θ	$\tan(\theta)$ is the slope of the inner aqueduct
ρ	volumetric mass of water, kg/m^3
Ω	area of the section of a derivation, m^2
Ω_{cit}	ground surface of the cistern, m^2

Subscript

A	at point A (see Fig. 8.16)
aq	in the inner aqueduct
B	at point B (see Fig. 8.16)
C	at point C (see Fig. 8.16)
cit	cistern at the end of the first derivation

Appendix 1: Methodology for the Characterization of the Third Hydraulic System

Global Description of the Flow in a Derivation

In Fig. 8.16, the first derivation is schematically presented. Three important points of the flow in the derivation are presented in Fig. 8.16. Point A is located in the connection between the aqueduct and the room of visit. Point B is located in the first pipe of the canalization, connected to the room of visit. Point C is located in the last pipe of the canalization, just before the entrance in the cistern.

The pressure at point A (P_A) is equal to the pressure in the bottom of the inner aqueduct, minus the loss of volumetric energy at the entrance in the derivation ($\rho V_A{}^2/4$, (Lencastre, 1995)) and minus the loss of pressure needed to create the kinetic energy at point A, $\rho V_A{}^2/2$. The pressure in the bottom of the aqueduct is simply equal to the atmospheric pressure (P_{atm}) plus the hydrostatic pressure created by the weight of water in the aqueduct ($\rho g h$). Therefore, the following equation can be written:

$$P_A = P_{atm} + \rho g h - \frac{3}{4}\rho V_A^2 \tag{8.3}$$

If the level of water in the cistern is below the end of the derivation, the pressure at point C is simply the atmospheric pressure:

$$P_C = P_{atm} \tag{8.4}$$

If the level of water in the cistern is above the end of the derivation, the pressure at point C is the atmospheric pressure plus a hydrostatic pressure created by the water in the cistern:

$$P_C = P_{atm} + \rho g(H - H_d) \tag{8.5}$$

where H is the height of water in the cistern and H_d is the height difference between the end of the canalization and the bottom of the cistern (see Fig. 8.13).

If the reference of the potential energy is fixed at point C, the volumetric energy of the flow at point C can be written as follows:

$$E_C = P_C + \frac{1}{2}\rho V_C^2 \tag{8.6}$$

where V_C is the velocity at point C.

If D_1 is the height difference between point C and point A, the volumetric energy of the flow at points A and B can be written as follows:

$$E_A = P_A + \frac{1}{2}\rho V_A^2 + \rho g D_1 \tag{8.7}$$

$$E_B = P_B + \frac{1}{2}\rho V_B^2 + \rho g D_1 \tag{8.8}$$

$E_A > E_B$ as there is an energy loss in the room of visit. The energy loss between two points in a flow is classically expressed as being proportional to the kinetic energy of the flow at the upstream point:

$$E_A - E_B + \frac{1}{2}C_{R1}\rho V_A^2 \tag{8.9}$$

where C_{R1} is called the energy drop coefficient of the flow between points A and B.

$E_B > E_C$ as there is an energy loss in the canalization and in the bend. The difference of volumetric energy between points B and C can be expressed as follows:

$$E_B - E_C = \frac{1}{2}L_1 C_{C1}\rho V_B^1 + \frac{1}{2}C_B\rho V_B^2 \tag{8.10}$$

where C_{C1} is the energy drop coefficient of the flow in 1 m of the canalization, L_1 is the length of the canalization and C_B is the energy drop coefficient of the flow in the bend.

If the flow rate in the derivation is written Q (kg/s), the following equations can be written:

$$Q = \rho\Omega_A V_A = \rho\Omega_B V_B = \rho\Omega_C V_C \tag{8.11}$$

where Ω_{A} is the area of the section of the connection between the room of visit and the aqueduct and $\Omega_{\mathrm{B}} = \Omega_{\mathrm{C}}$ is the area of the section of the pipes in the canalization.

When the level of water in the cistern is below the end of the derivation, the combination of Equations (8.3), (8.4), (8.6), (8.7), (8.8), (8.9), (8.10) and (8.11) yields:

$$Q = \rho\Omega_{\mathrm{B}} V_{\mathrm{B}} = \rho\Omega_{\mathrm{B}} \sqrt{2g \frac{h + D_{\mathrm{l}}}{1 + \frac{1}{2}\left(\frac{\Omega_{\mathrm{B}}}{\Omega_{\mathrm{A}}}\right)^2 + L_1 C_{\mathrm{C1}} + C_{\mathrm{B}} + C_{\mathrm{R1}}\left(\frac{\Omega_{\mathrm{B}}}{\Omega_{\mathrm{A}}}\right)^2}} \tag{8.12}$$

When the level of water in the cistern is above the end of the derivation, the combination of Equations (8.3), (8.5), (8.6), (8.7), (8.8), (8.9), (8.10) and (8.11) yields:

$$Q = \rho\Omega_{\mathrm{B}} V_{\mathrm{B}} = \rho\Omega_{\mathrm{B}} \sqrt{2g \frac{h + D_{\mathrm{l}} - (H - H_{\mathrm{d}})}{1 + \frac{1}{2}\left(\frac{\Omega_{\mathrm{B}}}{\Omega_{\mathrm{A}}}\right)^2 + L_1 C_{\mathrm{C1}} + C_{\mathrm{B}} + C_{\mathrm{R1}}\left(\frac{\Omega_{\mathrm{B}}}{\Omega_{\mathrm{A}}}\right)^2}} \tag{8.13}$$

A mass balance on the water in the cistern can be written as follows:

$$\rho\Omega_{\mathrm{cit}} \frac{dH}{dt} = Q = \begin{cases} \rho\Omega_{\mathrm{B}} \sqrt{2g \frac{h + D_{\mathrm{l}}}{1 + \frac{1}{2}\left(\frac{\Omega_{\mathrm{B}}}{\Omega_{\mathrm{A}}}\right)^2 + L_1 C_{\mathrm{C1}} + C_{\mathrm{B}} + C_{\mathrm{R1}}\left(\frac{\Omega_{\mathrm{B}}}{\Omega_{\mathrm{A}}}\right)^2}}, \text{when } H < H_{\mathrm{d}} \\ \rho\Omega_{\mathrm{B}} \sqrt{2g \frac{h + D_{\mathrm{l}} - (H - H_d)}{1 + \frac{1}{2}\left(\frac{\Omega_{\mathrm{B}}}{\Omega_{\mathrm{A}}}\right)^2 + L_1 C_{\mathrm{C1}} + C_{\mathrm{B}} + C_{\mathrm{R1}}\left(\frac{\Omega_{\mathrm{B}}}{\Omega_{\mathrm{A}}}\right)^2}}, \text{when } H < H_{\mathrm{d}} \end{cases} \tag{8.14}$$

where Ω_{cit} is the ground surface of the cistern.

Computational Fluid Dynamics of Turbulent Flows

The turbulent flow of isothermal and incompressible water can be described by the Reynolds average continuity and Navier Stokes equations (Kundu and Cohen, 2004):

$$\frac{\partial U_i}{\partial t} + U_j \frac{\partial U_i}{\partial x_j} = \frac{-1}{\rho}\frac{\partial P}{\partial x_i} - \frac{\partial z}{\partial x_i} + \frac{\partial}{\partial x_j}\left([\upsilon + \upsilon_t]\frac{\partial U_i}{\partial x_j}\right) \tag{8.15}$$

$$\frac{\partial U_j}{\partial x_j} = 0 \tag{8.16}$$

U_i is the component of the velocity along the x_i axis. P is the pressure and ν is the molecular kinematic viscosity of the liquid. ν_t is the turbulent viscosity (Pope, 2000).

In this work, k-ϵ models of turbulence are used. k is the turbulent kinetic energy and ϵ its dissipation rate. In k-ϵ models, ν_t is related to k and ϵ, and balance equations for k and ϵ complete Equations (8.15) and (8.16) (Pope, 2000; Versteeg and Malalasekera, 2007).

In most cases, Equations (8.15) and (8.16) cannot be solved analytically. Therefore, a numerical treatment of these equations has to be considered. In the finite volume technique (Versteeg and Malalasekera, 2007), the studied domain is meshed: it is divided into a set of small tetrahedral or hexahedral cells. Then, the conservation equations are integrated on each of these cells and finite difference schemes are applied. In this way, the initial system of partial differential equations can be transformed into a set of non-linear algebraic equations. The values of U_i at the centre of the cells are amongst the unknown of this set of equation. Finally, this system of algebraic equations is solved numerically. In this work, the finite volume technique implemented in Fluent 6.3 is used. The segregated double precision solver is used, with the SIMPLE velocity-pressure coupling algorithm. Default discretization scheme are used, except for the flow in canalizations, where power law discretization schemes are used for the velocity, k and ϵ. Default values of under relaxation factors are used. Geometries and meshes are generated in the software Gambit, and imported in Fluent.

Equations (8.15) and (8.16) have to be supplemented by adequate boundary conditions. When the flow is turbulent, a special care has to be given to the boundary conditions for the velocity on the walls. Near a wall, a boundary layer is developed. Let us consider a cell adjacent to a wall. y_{p} is the distance between its centre and the wall. The velocity magnitude at the centre of the cell is written U_{p}. The turbulent kinetic energy at the centre of the cell is written k_{p} and its dissipation rate is written ϵ_{p}. The boundary layer is divided into several sub-layers. Within a given range of y_{p}, there is a logarithmic relationship between U_{p} and y_{p} (Pope, 2000):

$$U_{\mathrm{p}} = \frac{1}{\alpha} \ln(\beta y_{\mathrm{p}}) - \Delta B \tag{8.17}$$

where ΔB quantifies the reduction of the velocity due to the roughness of the wall. ΔB can be well correlated to the physical roughness K. Such a correlation is implemented in Fluent 6.3.

In the k-ϵ models, the boundary condition for k imposed at walls is (Versteeg and Malalasekera, 2007):

$$\frac{\partial k}{\partial n} = 0 \tag{8.18}$$

where n is a direction perpendicular to the wall.

The turbulent kinetic energy dissipation rate at the centre of a cell adjacent to a wall is evaluated as follows (Versteeg and Malalasekera, 2007):

$$\varepsilon_{\mathrm{p}} = \frac{\lambda k_{\mathrm{p}}^{3/2}}{y_{\mathrm{p}}} \tag{8.19}$$

The coefficients α and β depend on k_{p}, U_{p} and y_{p}, while λ is a constant.

In this work, Equations (8.17), (8.18) and (8.19) are used as boundary conditions at walls. Therefore, in the simulations presented, most of the cells adjacent to walls have a size such that their y_p lies in the logarithmic sub-layer. As the velocity of the fluid near the wall increases, this sub-layer shrinks and heads towards the wall. Hence, as the flow rate increases, the mesh usually has to be refined.

Conditions used at other types of boundary conditions can be found in (Versteeg and Malalasekera, 2007), as well as a complete description of Fluent 6.3.

References

Bailhache M (1979) L'évolution du débit des aqueducs gallo-romains. Dossiers de l'Archéologie 38:82–87

Balty J (1987) Problèmes de l'eau à Apamée de Syrie. Les Travaux de la Maison de l'Orient 14:9–23

Balty J (2000) Claudia Apamea. Données nouvelles sur la topographie et l'histoire d'Apamée. Comptes rendus des séances de l'Académie des Inscriptions et Belles – Lettres janmar:459–486

Chanson H (2002) Certains aspects de la conception hydraulique des aqueducs romains. La Houille Blanche 6/7:43–57

Evans H (1994) Water distribution in ancient Rome: The evidence of Frontinus. The University of Michigan Press, Ann Arbor

Herschel C (1973) The two books on the water supply of the city of Rome of Sextus Julius Frontinus. New England Water Works Association, Boston

Kundu P, Cohen I (2004) Fluid mechanics. Academic Press, London

Lencastre A (1995) Hydraulique générale. Eyrolles, Paris

Morgan M (1960) Vitruvius. The ten books on architecture. Dover Publications, New York

Ortloff C, Crouch P (2001) The urban water supply and distribution system of the Ionian city of Ephesos. J Archaeological Sci 28:843–860

Pope S (2000) Turbulent flows. Cambridge University Press, Cambridge

Versteeg H, Malalasekera W (2007) An introduction to Computational Fluid Dynamics: The finite volume method approach, 2nd edn. Prentice Hall

Viollet P (2000) L'hydraulique dans les civilisations anciennes. Presse de l'Ecole des Ponts et Chaussées, Paris

Viviers D, Vokaer A (2009) Quartier Nord-Est. Observations sur le système d'adduction d'eau de la ville. In: Proceedings of the IVème Colloque Apamée de Syrie. Bilan des recherches archéologiques 1980–2004, Presses de l'Université Libre de Bruxelles, Bruxelles (in press)

Whitaker S (1968) Introduction to fluid mechanics. Prentice Hall, New York

Chapter 9
Water Technology in the Ancient American Societies

Larry W. Mays and Yuri Gorokhovich

9.1 Introduction

Many ancient civilizations in the Americas developed water technologies during the same times that water technology was advancing in other parts of the world. This chapter will address water technologies of societies of the pre-Columbian empires in the southwestern United States, Mesoamerica, and the Inca in South America. These locations are shown in Fig. 9.1. In the southwestern U.S. we will discuss water technologies of the Hohokam and the Anasazi. In Mesoamerica we will discuss the water technologies of the Teotihuacan, the Chaco Anasazi, the Xochicalco, the Maya and the Aztec. Finally the Inca water systems of Machu Picchu, Tipon, Pisac, Tambo Machay and Puca Pucara are discussed. For further reading on the failure of some of these ancient civilizations, Mays (2007a, b) discusses the water sustainability of ancient civilizations in Mesoamerica and the American Southwest.

9.2 American Southwest

Three major cultures – the Chaco Anasazi, the Hohokam, and the Mogollon- existed in the American southwest during the late pre-contact period. Figure 9.2 shows the extent of the Chaco Anasazi and the Hohokam. The concept of prehistoric regional systems has been used to describe these cultures (Crown and Judge, 1991). The Hohokam and Chaco Anasazi regional systems have received particular attention as two of the most important. The extent of the Hohokam regional system has been defined by ball courts and material culture, and the Chaco regional system has been defined by roads and other architectural criteria. The American southwest is a difficult and fragile environment consisting of arid and semi-arid lands. The Chaco and

L.W. Mays (✉)
School of Sustainable Engineering and the Built Environment, Arizona State University, Tempe, AZ 85287-5306, USA
e-mail: mays@asu.edu

L.W. Mays (ed.), *Ancient Water Technologies*, DOI 10.1007/978-90-481-8632-7_9,

Fig. 9.1 Pre-Columbian Empires in North America, Mesoamerica, and South America

Hohokam systems evolved in quite different environments as will be explained. The Hohokam system had the benefits of the Sonoran desert and the Gila River. The Chaco system was located in the harsh and uncertain environment of the San Juan Basin. Each of these occupied a distinctive ecological niche within the southwestern environment, and as a consequence their infrastructures significantly differed.

9.2.1 The Hohokam (A.D. 1–1450)

Hohokam, translated as "the people who vanished", is the name given to their prehistoric predecessors by the present-day Pima Indians. They lived in south-central Arizona and northern Mexico from approximately A.D. 1–1450 (Andrews and Bostwick, 2000). Floodwater farming was practiced using floodplain inundation (overbank flooding) and ak-chin farming (consisting of capturing rainfall runoff in the fields). The word ak-chin is derived from the Papago word meaning "mouth of a wash" (Masse, 1991). Dry farming was also practiced by the Hohokam, which consisted of using agricultural techniques that were constructed to utilize direct rainfall

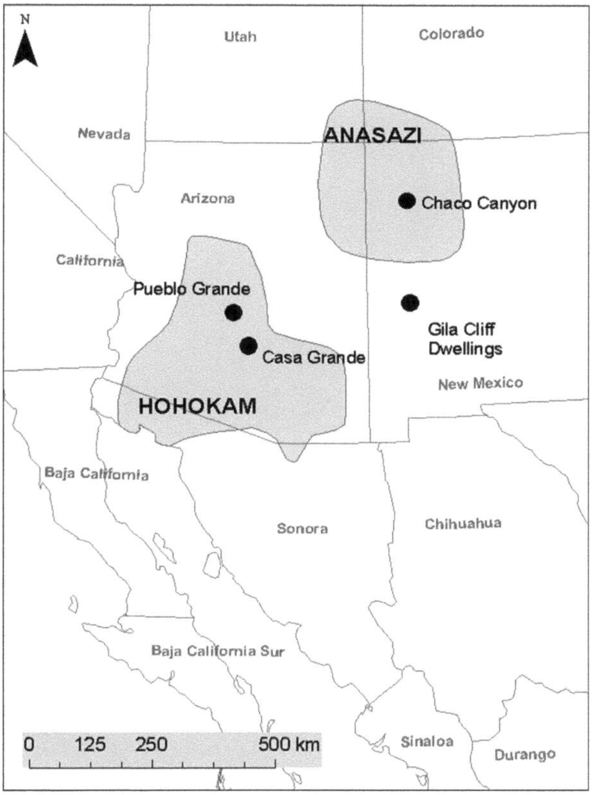

Fig. 9.2 Ancient cultures in the American Southwest

or to divert rainfall-runoff short distances to fields (Masse, 1991). This method was practiced in the less dry part of the Hohokam regional system.

The Hohokam built the most complex irrigation system in the desert lowlands of the Salt-Gila River Basin, Arizona in and around the present day Phoenix, Arizona. They built more than 483 kilometers (km) of major canals and over 1,126 km of distribution canals in the Salt River Valley, which have been identified, see Fig. 9.3. The Hohokam civilization started in the Valley somewhere between 300 B.C. and A.D. 1 (see Crown and Judge, 1991) and extended to A.D. 1450 (Lister and Lister, 1983). Figure 9.4 shows Hohokam canals at Park of the Canals in Mesa, Arizona.

A schematic representation of the major components of a Hohokam irrigation system is shown in Fig. 9.5. Hohokam irrigation systems consisted of three basic types of canals: main canals, distribution canals, and field laterals. Main canals as shown in Fig. 9.5 extend from the canal intake at the river to the first major junction where the canal size is significantly reduced in size. Distribution canals branch off either from the main canals or from other distribution canals and supply

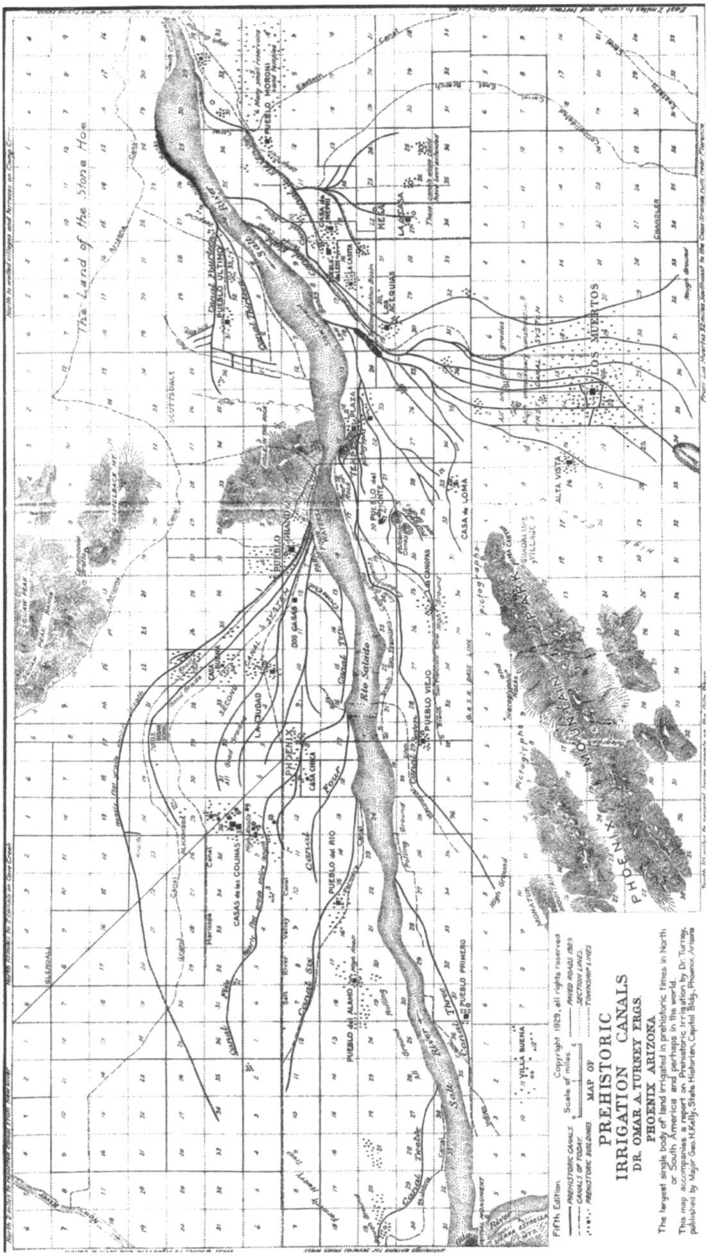

Fig. 9.3 Hohokam canal system in Salt River Valley, Turney (1929)

(a) (b)

Fig. 9.4 Hohokam Canals at Park of the Canals in Mesa, Arizona (photos copyright by L.W. Mays)

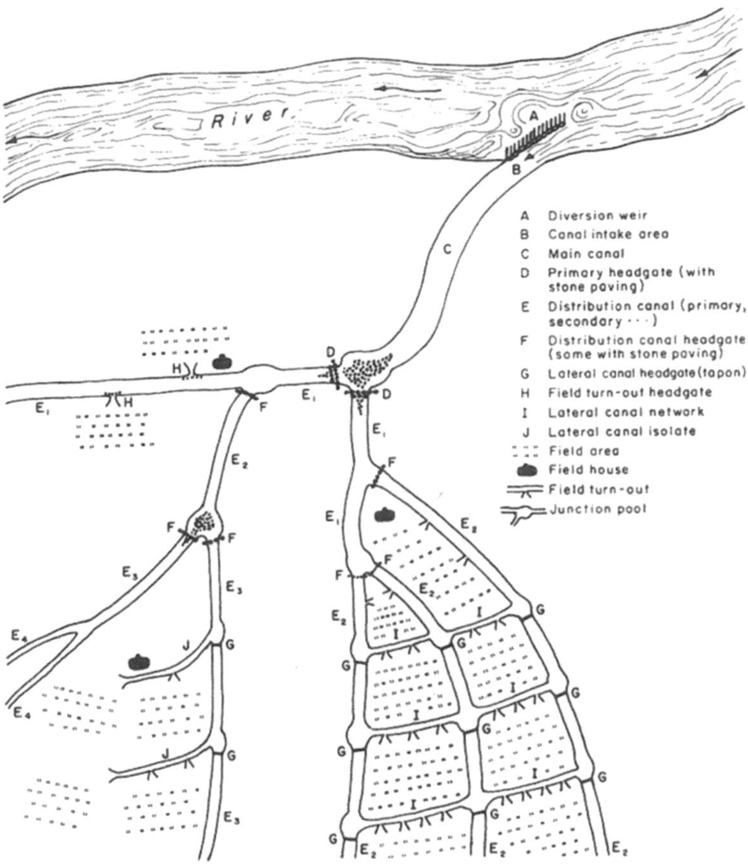

Fig. 9.5 Components of Hohokam irrigation system. Masse (1991)

water to the field laterals which are the terminal segment of the system. Two different types of laterals, the lateral network and the lateral canal isolate, were utilized as shown in Fig. 9.5. Lateral networks refer to those that are arranged in parallel sets and are coupled with multiple distribution canals to create webs whereas the lateral canal isolate did not necessarily operate in tandem with other laterals (Masse, 1991).

There were other water conveyance features of the irrigation systems including canal junctions, canal head gates, junction pool areas for water control, erosion control structures made of stone paving, and tapon post and mat structures to raise canal water levels into field laterals and turnouts. Canal head gates constructed of logs and brush were utilized to regulate the velocity and discharge in

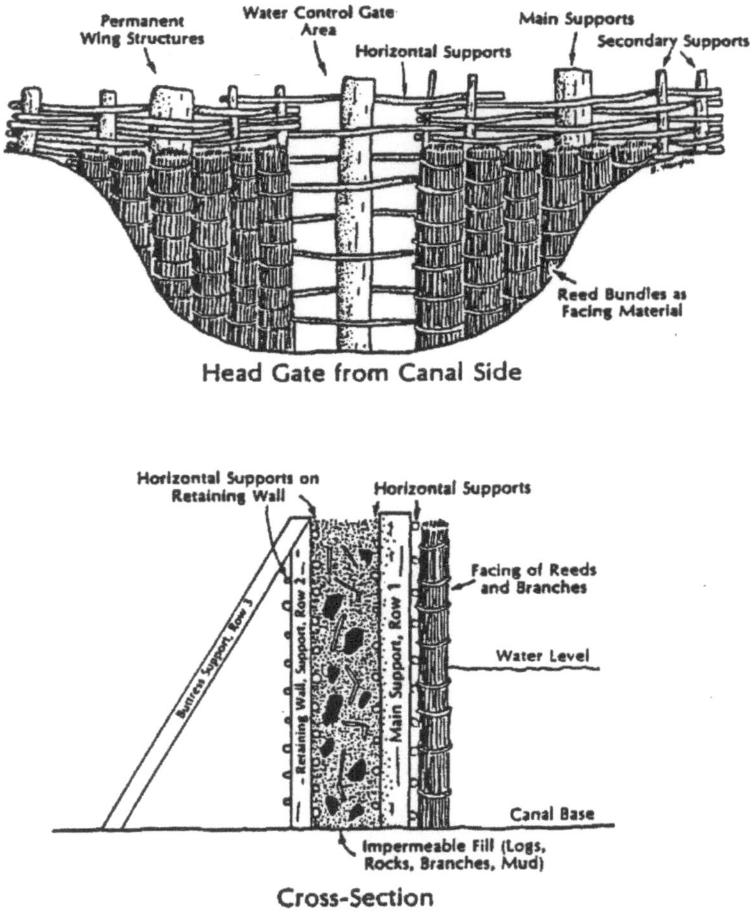

Fig. 9.6 Suggested configuration of a Hohokam canal head gate (Ackerly et al., 1987)

the canals. Figure 9.6 shows a suggested configuration of Hohokam canal head gates, which were constructed of vertical log posts that were set into the canal bed. A watertight barrier was created by interweaving brush and bundles of reeds into the posts.

The irrigation system lacked some critical features such as permanent dam structures, drop structures, structures to prevent silting, and the use of canal linings. Canal linings would have retarded water leakage to prevent water logging, minimize damage from flood flows, and the salinization of the fields. We must emphasize that the Hohokam irrigation canal system was built with little technology other than the use of stone tools, sharpened sticks, and carrying baskets.

In A.D. 899 a flood caused decentralization and widespread population movement of the Hohokams from the Salt-Gila River Basin to areas where they had to rely upon dry farming. The dry farming provided a more secure subsistence base. Eventual collapse of the Hohokam regional system resulted from a combination of several factors. These included flooding in the 1080's, hydrologic degradation in the early 1100's, and larger communities forcibly recruiting labor or levying tribute from surrounding populations (Crown and Judge, 1991). In 1358 a major flood ultimately destroyed the canal networks, resulting in the depopulation of the Hohokam area. Culturally drained the Hohokam faced obliteration about 1450. Parts of the irrigation system had been in service for almost fifteen hundred years and most likely were in severe disrepair, canals silted requiring extensive maintenance, and problems with salt. See Haury (1978), Hunt, et al. (2005), Masse (1981), and Woodbury (1960) for further information.

9.2.2 The Chaco Anasazi (A.D. 600–1200)

In the high deserts of the Colorado Plateau (see Fig. 9.2), the Anasazi (a Dine' (Navajo) word meaning "enemy ancestors"), also called the "ancient ones", had their homeland. When the first people arrived in Chaco Canyon, there were abundant trees, a high groundwater table, and level floodplains without arroyos. This was most likely an ideal environment (conditions) for agriculture in this area. Chaco is beautiful with four distant mountain ranges: the San Juan Mountains to the north; the Jemez Mountains to the east; the Chuska Mountains to the west; and the Zuni Mountains to the south.

The first Anasazi settlers, also called basket makers, arrived in Mesa Verdi around A.D. 600. They entered the early Pueblo phase (A.D. 700–900) which was the time they transitioned from pit houses to surface dwellings, evidenced by their dramatic adobe dwellings, or pueblos. Chaco Canyon was the center of the Anasazi civilization, with many large pueblos probably serving as administrative and ceremonial centers for a widespread population of the Chaco regional system. Also of particular note is the extensive road system, built by a people who did not rely on either wheeled vehicles or draft animals. The longest and best-defined roads (constructed between A.D. 1075 and 1140) extended over 80.5 km in length. The rise and fall of the Chacoan civilization was from A.D. 600 to 1200, with the peak decade being A.D. 1110–1120.

Chaco Canyon is situated in the San Juan Basin in northwestern New Mexico as shown in Fig. 9.2. The basin has limited surface water, most of which is discharged from ephemeral washes and arroyos. Figure 9.7 illustrates the method of collecting and diverting runoff throughout Chaco Canyon. The water, collected from the side canyon that drained from the top of the upper mesa was diverted into canals by either an earthen or a masonry dam near the mouth of the side canyon (Vivian, 1990). These canals averaged 4.5 m in width and 1.4 m in depth; some were lined with stone slabs and others were bordered by masonry walls. The canals ended at a

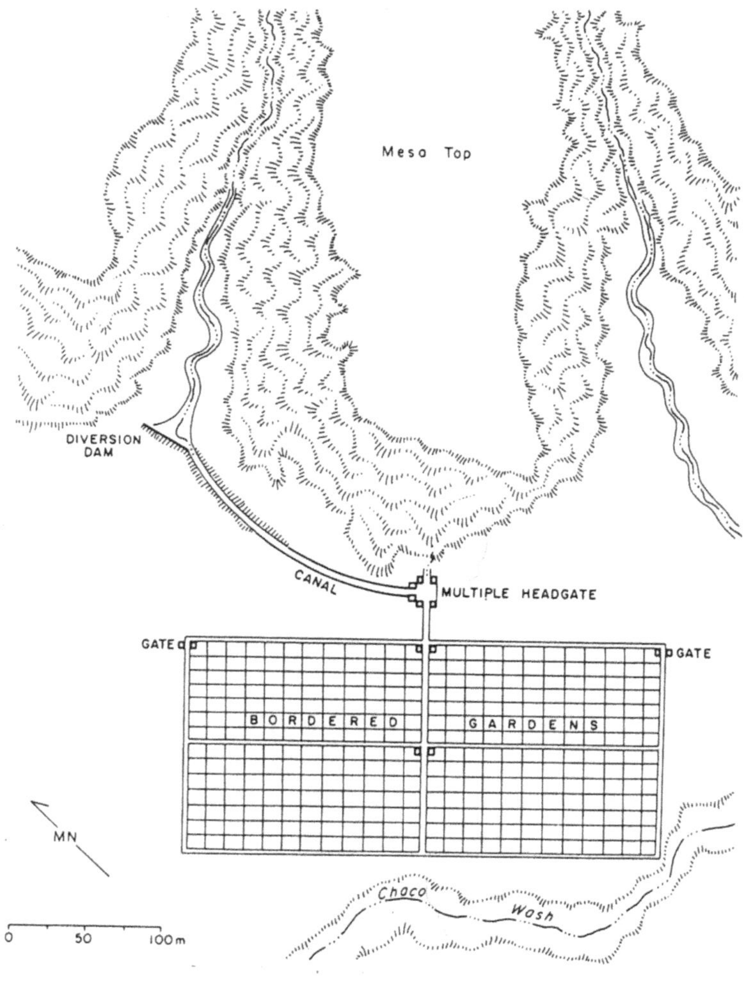

Fig. 9.7 Water-control System in Chaco Canyon. Vivian (1974)

masonry head gate, where water was then diverted to the fields in small ditches or to overflow ponds and small reservoirs.

9.3 Mesoamerica

Mesoamerica includes Mexico and the northern Central America. The earliest Mesoamerican civilization, the Olmecs, evolved sometime before 1000 B.C. along the Gulf of Mexico. After about 800 B.C. the Mesoamerican civilization exerted social and religious influence in an area extending from the Valley of Mexico to modern El Salvador. Many Mesoamerican civilizations developed and failed for various reasons. The period or era from about 150 A.D. to 900 A.D. (called the Classic) was the most remarkable in the development of Mesoamerica (Coe, 1994). During the Classic Period the people of Mexico and the Mayan area (see Fig. 9.1) built civilizations comparable with advanced civilizations in other parts of the world. In Mesoamerica the ancient urban civilizations developed in arid highlands where irrigation (hydraulic) agriculture allowed high population densities. In the tropical lowlands, however, there was a dependence on slash-and-burn (milpa) agriculture which kept the bulk of the population scattered in small hamlets. Sanders and Price (1968) suggest that the non-urban lowland civilization resulted from responses to pressures set up by the hydraulic, urban civilization. Teotihuacan (City of the Gods) in Mexico is the earliest example of highland urbanism.

9.3.1 Teotihuacan Empire (300–600 A.D.)

Teotihuacan (see Fig. 9.1) was a very impressive civilization which evolved about twenty-five miles north of Mexico City. Prior to 300 B.C., Teotihuacan valley had a small population spread over the valley and was the dominant urban center in Mesoamerica throughout the classic period. By A.D. 100 Teotihuacan covered an area of 12 km^2 which has been linked to the development of so-called hydraulic agriculture (Haviland, 1970). The urban area expanded in size, there was an increased socio-economic diversity, and an expanding political influence. At its height, around 600 A.D., Teotihuacan was fully urban with a population of approximately 85,000 people and covering an area of 19 km^2 (Haviland, 1970). Others (such as Million, 1983) have estimated that the maximum population was approximately 125,000 during the Xolalpan phase. Teotihuacan was the largest urban center of the time in Mesoamerica.

Around 300 B.C. the use of canals for irrigation rapidly spread throughout the central highland basin, the location of Teotihuacan (Doolittle, 1990). South of Teotihuacan near Amanalco, Texcoco east of the Basin of Mexico), irrigation would have consisted of diverting water from shallow spring-fed streams into simple irrigation canals and then onto fields only a few meters away. Flood water systems also were used. Northeast of Teotihuacan, south of Otumba, a series of ancient irrigation canals (dating between 300 and 100 B.C.) were excavated. Other canals in the same area date to A.D. 900–1600. Evidence also exists of canals that were built between A.D. 200 and 800 near Teotihuacan. One of these was the first confirmed relocation of a natural stream. The reader can refer to Doolittle (1990) for further information on these canals.

9.3.2 Xochicalco (A.D. 650–900)

After the disintegration of Teotihuacan's empire in the 7th century A.D., foreigners from the Gulf Coast lowlands and the Yucatan Peninsula appeared in central Mexico. Cacaxtla and Xochicalco, both of Mayan influence, are two regional centers that became important with the disappearance of Teotihuacan. Xochicalco (in the place of the house of flowers), was located on hill top approximately 38 km from modern day Cuernavaca, Mexico, and became one of the great Mesoamerican cities in the late classic period (A.D. 650–900). Figure 9.8a and b show scenes of the city, which was well thought out with terraces, streets, plazas, and buildings. Despite the very Mayan influences, the predominant style and architecture is that of Teotihuacan. There were no rivers or streams or wells to obtain water. Water was collected in the large plaza area and conveyed into cisterns such as the one shown in Fig. 9.8c. From the cisterns water was conveyed to other areas of the city using terracotta pipe as shown in Fig. 9.8d.

9.3.3 The Maya

The ancient Maya (Coe, 1993) lived in a vast area covering parts of present-day Guatemala, Mexico, Belize, and the western areas of Honduras and El Salvador, see Figs. 9.1 and 9.9. Mayans settled in the last millennium B.C. and their civilization flourished until around 870 A.D. The environment that the Mayans lived in was less fragile than that of the semi-arid lands where the Anasazi and Hohokam lived. Tikal was one of the largest lowland Maya centers, located some 300 km north of present day Guatemala City. The city was located in a rain forest setting with a present day average annual rainfall of 135 cm. The urbanism of Tikal was not because of irrigated (hydraulic) agriculture. A number of artificial reservoirs were built in Tikal, which became more and more important as the population increased.

The Mayas settled in the lowlands of the Yucatan Peninsula and the neighboring coastal regions. The large aquifer under this area is in an extensive, porous limestone layer (Karst terrain), which allows tropical rainfall to percolate down to the aquifer. Because of this and the fact that few rivers or streams exist in the area, surface water is scarce. One important water supply source for the Maya, particularly in the north, was the underground caves (see Fig. 9.10) called cenotes (se-NO-tes), which also had religious significance (portals to the underworld where they journeyed after death to meet the gods and ancestors). In Yucatan there are over 2,200 identified

Fig. 9.8 Photos of Xochicalco (a) View showing Xochicalco on hilltop, (b) View from Xochicalco, (c) Cistern, Color version available in Appendix (d) Water pipes Color version available in Appendix (photos copyright by L.W. Mays)

and mapped cenotes. In the south the depths to the water table were too great for cenotes.

Natural surface depressions were lined to reduce seepage losses and were used as reservoirs. Another source was water that collected when soil was removed for house construction in depressions called aguados. The Maya also constructed cisterns called chultans in limestone rock under buildings and ceremonial plazas. The chultans were bottle-shaped underground cisterns that were dug in limestone

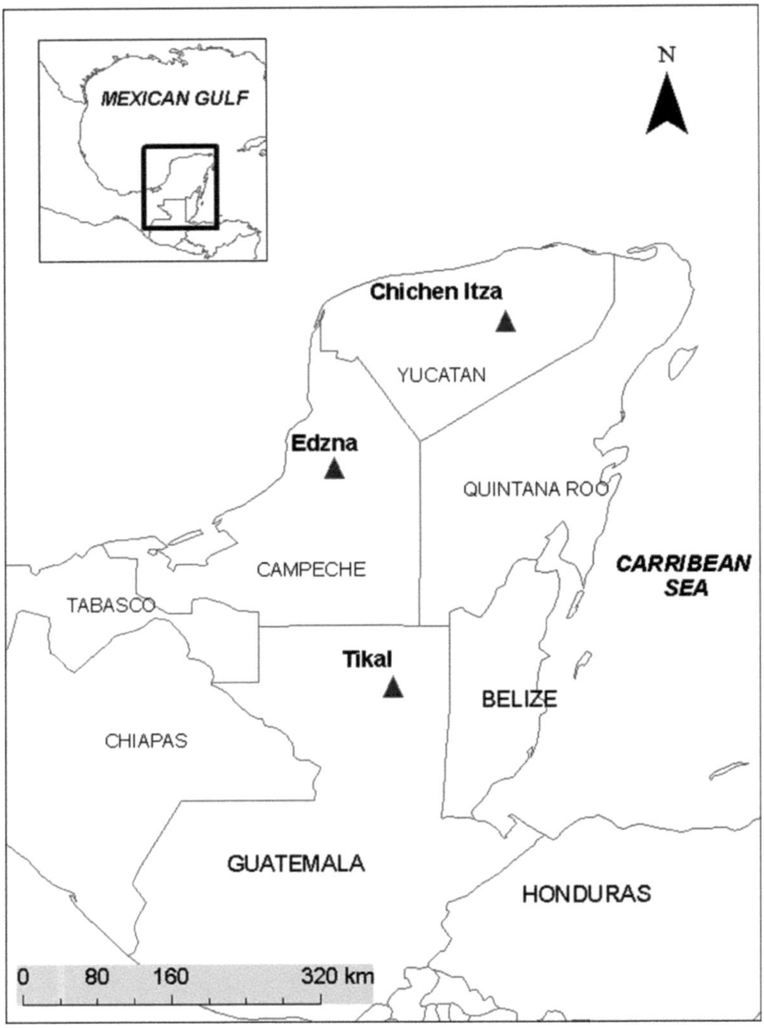

Fig. 9.9 Map of Maya area

bedrock and plastered with cement. Drainage systems were developed to divert surface runoff from buildings and courtyards or plazas into the chultans. In the lowlands the Maya typically used one or more of these methods for obtaining and storing water supplies (Matheny et al., 1983).

In the central Campeche valley in the Yucatan Peninsula is the location of Edzna (see Fig. 9.9). During the Late Classic Period an extensive hydraulic system was constructed at Edzna. This system consisted of more than 20 km of canals and

Fig. 9.10 Sacred cenote at Chichen Itza (which means mouth of the well of the Itzas). The word cenote is derived from *tz'onot*, the Maya term for the natural sinkholes. This cenote, which measures about 50 m form north to south and 60 m from east to west, was used for sacrifices of young men and women, warriors and even children to keep alive the prophecy that all would live again. Shown at the left are the remains of a building once used as a steam bath, or *temezcal*, to purify those to be sacrificed. Those sacrificed were tossed from a platform that jutted out over the edge of the cenote (photo copyright by L.W. Mays)

an extensive system of reservoirs (Matheny, 1976). The system was operational by the time of Christ, as suggested by excavations and analysis of artifacts (Matheny, 1976). Figure 9.11 is a map showing the various hydraulic features at Edzna. The seven large canals (numbered 1 through 7 in Fig. 9.11) averaged 40 m wide and ranged from 0.6 to 1.5 km long and the two smaller canals (canals 8 and 9) are shown in Fig. 9.11 (Matheny, 1976). The reservoirs in the northern part are mostly independent but some are connected to the canals. There are about 25 large reservoirs accompanied by mounds and numerous small reservoirs each accompanied by one or more house mounds (Matheny, 1976). The exact special purposes of the hydraulic system are not known; however, it is safe to say that the rainwater was used for drinking. The modern Maya use the water in the ancient channels for drinking and bathing until about February when the water level is low and the channels are full of aquatic life and unfit for drinking (Matheny, 1976) Normal dry season is from January through May. The Maya of the Late Classic Period chose not to dig wells and only twelve chultans have been found at the site (Matheny, 1976). It is possible that the water system of Edzna is an example of how the Maya people of the lowlands controlled and used water.

Rainfall varies significantly from the north (46 cm/year) to the south (254 cm/year) of the Yucatan Peninsula. The soils are also deeper in the southern part resulting in more productive agriculture and consequently supported more people. Rainfall was very unpredictable, resulting in droughts that destroyed crops. Ironically though, the water problems were more severe in the wetter southern part. Ground elevations increased from the north to the south causing the depths down to the water table to be greater in the south.

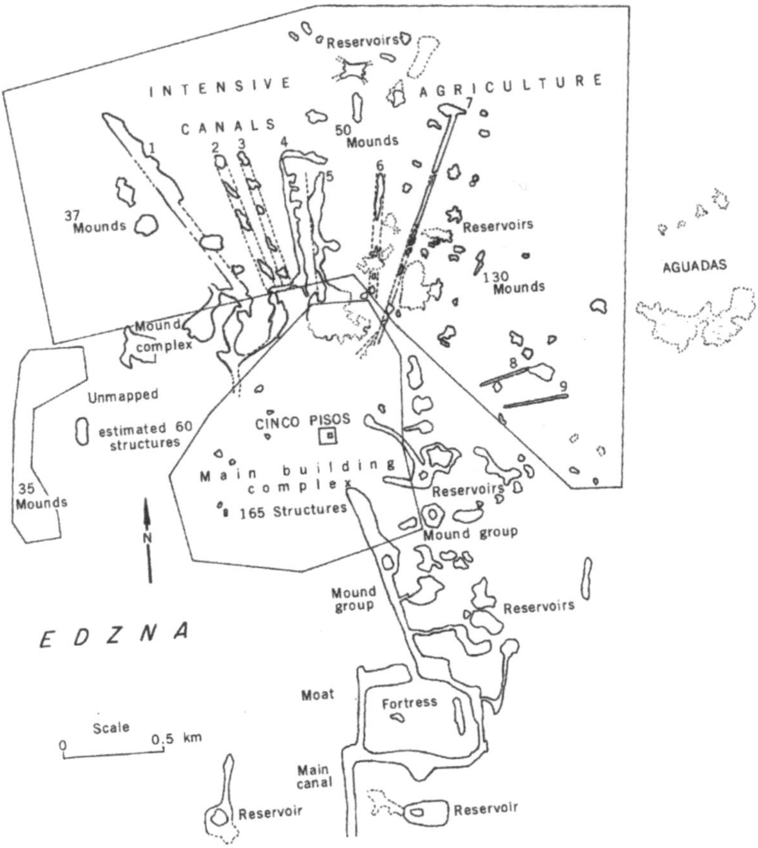

Fig. 9.11 Map of water system at Edzna. The outlined areas represent water-soaked canals and reservoirs determined from infrared aerial photographs and wet season mapping. Canals are aligned toward the center of the site. Canals number 1 through 7 are the large canals and 8 and 9 are smaller ones. (Matheny, 1976)

9.3.4 The Aztec Empire (A.D. 1150–1519)

Starting in the 12th century the nucleus of the Aztec Empire was the Valley (or Basin) of Mexico. Figure 9.1 shows approximate boundaries of the Aztec Empire. The city of Tenochtitlan, located on a reclaimed island in the saline Lake Texcoco, was the capital of the Aztec Empire, also known as the Aztec Triple Alliance. The Aztec Empire influence expanded its political hegemony far beyond the Valley of Mexico, conquering other city states throughout Mesoamerica. They overcame a lack of conventional farming land by building floating fields, *chinampas*, in Lake Texcoco (Woolf, 2005). In shallow areas of the lake the *chinampa* were built up using earth and plants. These artificial fields were held in place using wooden poles driven into the lake bed. In deeper areas of the lake reed beds, anchored to the lake bottom, were filled using earth. Obviously such a system was subject to flooding and could be surrounded easily by enemies. The chinampa cultivation may have originated with the Teotihuacan in the Early Classic period (Coe, 1994).

During the Golden Age (A.D. 1200–1520) in Mexico advances were being made in water technology, particularly in the development of aqueduct systems for irrigation and domestic water supply. Tenochtitlán made use of an elaborate system of aqueducts to carry spring water from higher elevations of the southern portion of the Basin of Mexico. "The aqueducts built in the Basin of Mexico during the pre-Hispanic times were so numerous and outstanding that they were one of the first indigenous engineering accomplishments that the Spaniards noted," (Doolittle, 1990). Figure 9.12 is a map of the western section of Lake Texcoco with Tenochtitlan showing the aqueducts from Coyohuacan and Chapultepec to Tenochtitlan. The one aqueduct given the most attention by scholarly researchers is the aqueduct that was built from Chapultepec Hill to Tenochtitlan located on an island. According to legend the source of water for the aqueduct gushed out

Fig. 9.12 Map of western section of Lake Texcoco, Tenochtitlan, and aqueducts (after Sanders, 1981 as presented in Doolittle, 1990)

from under the base of a large rock (Brundage, 1979). The place was sanctified by the Aztecs posting a priest of Tlaloc there (Brundage, 1979). The aqueduct from Chapultepec Hill to Tenochtitlan was rebuilt at least once. Construction began on the first aqueduct in A.D. 1418 (Bribiesca-Castrejon, 1958).

Details of the first aqueduct portion that traversed the water of Lake Texcoco are in Fig. 9.13. A chain of small islands were built with aqueduct troughs (of hallowed out split tree trunks) connecting each island. Also connecting each island were wooden planks laid next to the aqueduct which served as pedestrian crossings as shown in Fig. 9.13. Doolittle (1990) provides an excellent description of the island construction process. Reed mats (rafts) were weaved of unknown lengths with widths between seven and eight meters. These mats were floated into place and then anchored with stakes driven deep into the lake bed. Then rocks, mud and sod were loaded onto the mats until they sank to the lake bottom creating the island. The islands were constructed with a spacing of three to four meters apart creating a chain

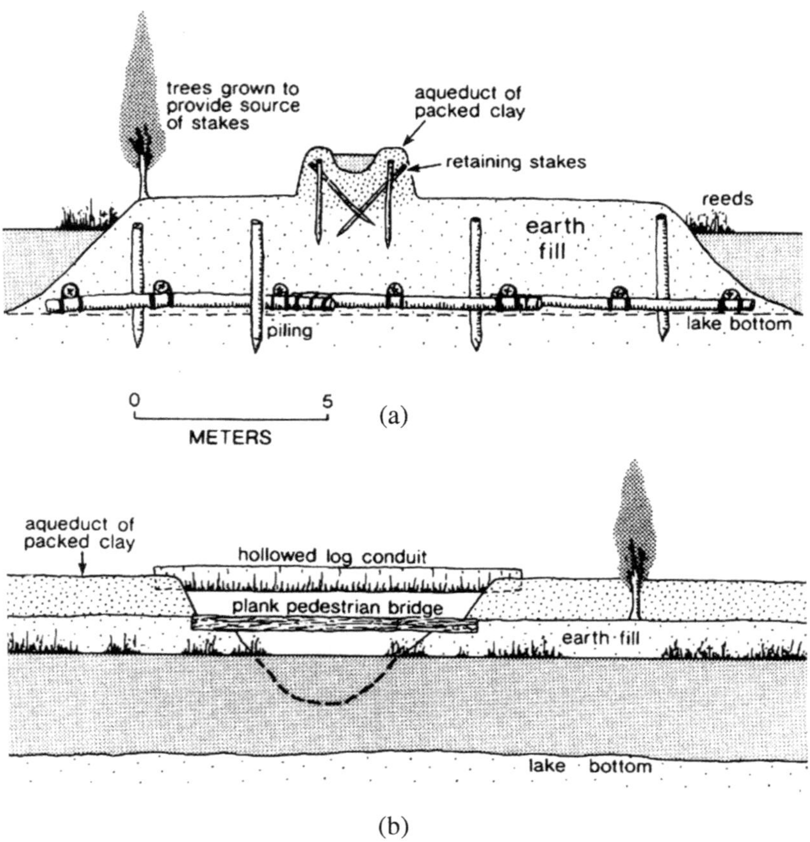

Fig. 9.13 First aqueduct from Chapultepec to Tenochtitlan (after Bribiesca Castrejon, 1958 as present in Doolittle, 1990)

of islands. On each island an aqueduct of packed clay was built over the length of the island as shown in Fig. 9.13. The aqueduct was approximately one meter high and 1.5–2 m wide (Doolittle, 1990). The first aqueduct was destroyed by a flood in A.D. 1449.

By A.D. 1465 a new aqueduct was rebuilt to replace the first one with a chain of islands connected with hollowed out logs. Possibly remains of the old islands were used. The new islands were much larger than the original ones, with widths of 10–12 m and lengths of tens of meters and were higher above the water level (Doolittle, 1990). Details of the rebuilt aqueduct from Chapultepec Hill to Tenochtitlan are shown in Fig. 9.14. The new aqueduct super structures were masonry with a lime-based mortar and dimensions of 1.6 m high and 2.5–3.0 m top width, with a wider width at the bottom. The trapezoidal shaped superstructures were built on a foundation of mixed with lime and small stones. The foundation was also supported using pilings as shown in Fig. 9.14. The new aqueduct consisted of two conduits, which were mortared as the rest of the superstructure. Each conduit was trapezoidal with a 75 centimeter (cm) depth, and 30 cm width at the bottom and 60 cm width at the top.

The use of the water once it reached Tenochtitlan has been unclear with many scholars (including Bribiesca-Castrejon, 1958) assuming it was distributed through small canals for domestic uses. Others such as Parlerm (1955) feel that more water than needed was delivered for domestic uses and the bulk was used for agricultural purposes, particularly keeping in mind that two aqueducts existed. After the Aztec Empire was conquered in 1520, the Spaniards rebuilt the aqueducts and continued to use the spring water until the mid 1850 s. In 1846 the discovery of potable artesian ground water started extensive drilling of the aquifer.

Mesoamericans had a large number of fertility gods and goddesses, of which the rain god was among the most senior. The Aztecs were faithful worshippers of the rain god, Tlaloc, whose cult dates back as far as the Olmec civilization (Woolf, 2005). Tlaloc was honored with sacrifices in the form of blood and other offerings.

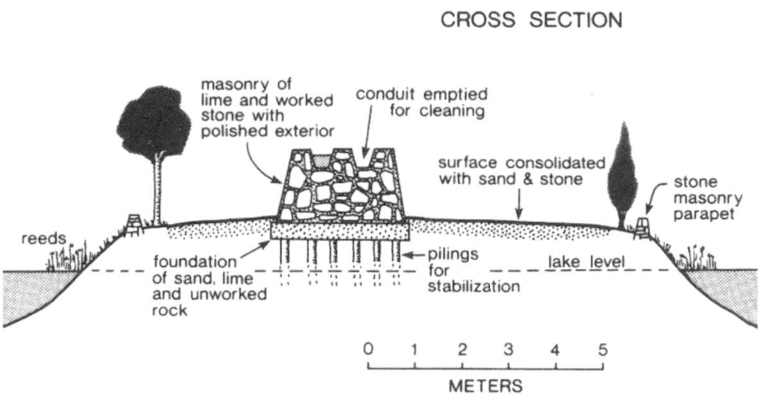

Fig. 9.14 Rebuilt aqueduct from Chapultepec to Tenochtitlan (after Bribiesca Castrejon, 1958 as present in Doolittle, 1990)

He stored rainwater in four huge jars, which he kept in the north, south, east, and west, and from the eastern jar he sent lifegiving rains, and from others storms and droughts (Woolf, 2005).

9.4 The Inca

People of the Inca civilization emerged from fragmented independent societies by A.D. 1000 (D'Altroy, 2003). The small city-state of Cuzco became the capital under the rule of Pachacuti Inka Yupanki and represented a center of Tawantinsuyu (The Four Parts Together), i.e. the Inca Empire. Figure 9.1 shows the approximate extent of the total Inca Empire, including parts of modern day Columbia, Ecuador, Peru, Bolivia, Chile and Argentina. Incas had developed water technologies to supply water for irrigation and domestic uses. Use of water for irrigation and religious purposes is mostly known from historical sites of Machu Picchu, Tambo Machay, Tipon, Puca Pucara, Cusichaca, Huanuco, Chinchero and Pisac. Three large Inca historical sites, Machu Picchu, Pisac and Tipon, locations shown in Fig. 9.15, are vivid examples of these technologies on a grand scale. For further reading on the Inca refer to D'Altroy (2003), Bauer (1998, 1992) and Hyslop (1990).

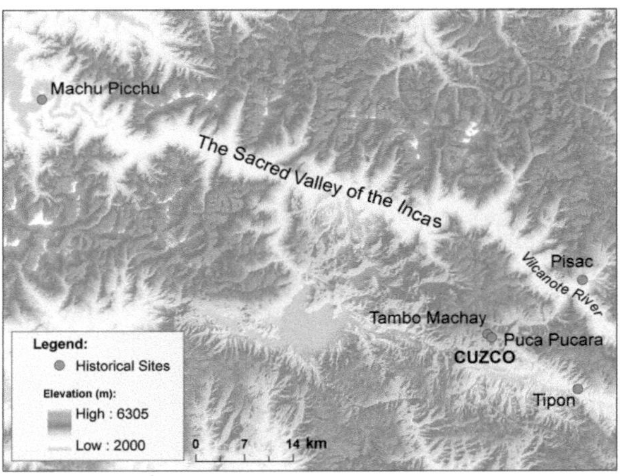

Fig. 9.15 General map of the Machu Picchu and Cuzco areas with historical sites related to water use and worshiping

Machu Picchu, one of the most familiar symbols of the Inca Empire, was built around 1450 A.D. but was abandoned as an official site for the Inca rulers a hundred years later, at the time of the Spanish conquest of the empire. Machu Picchu, located 2,430 m above mean sea level, is situated on a mountain ridge above the Urubamba Valley in Peru, about 80 km northwest of Cuzco. The city was built in the classical Inca style using polished dry stones. Water was supplied by a canal and a series of

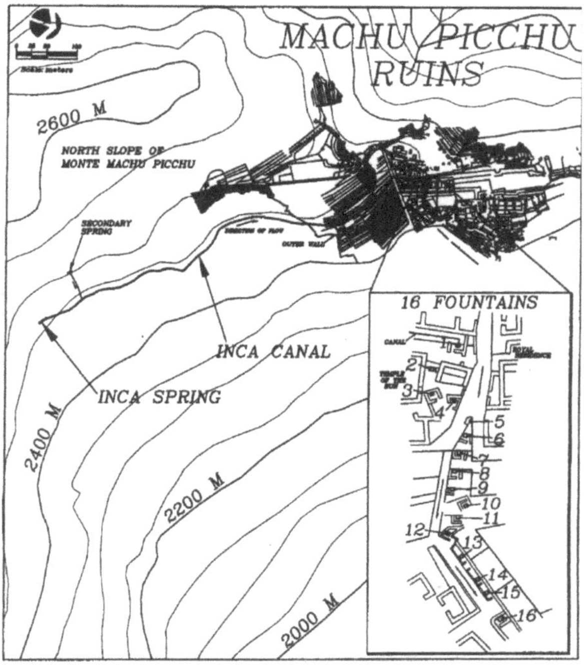

Fig. 9.16 Site map of Machu Picchu showing Inca canal and 16 domestic water supply fountains (Wright et al., 1997)

16 domestic fountains filled by a primary and a secondary springs (Fig. 9.16). The fountains are located in a hydraulic series over a distance of 55 m with a vertical drop of 26 m (Wright and Zegarra, 2000). Each fountain is unique, but all have an enclosure for privacy, an approach channel, a stone basin at the bottom, and an orifice outlet that could be plugged. Wright et al. (1997) and Wright and Zegarra (2000) discuss the locations of the canal and fountains.

Wright and Zegarra (1999, 2000) described the drainage system of Machu Picchu. The following discussion is based upon their description. This system consisted of several components including a centralized main drain separating the agricultural system from the urban drainage system. The agricultural drainage system included terrace surface drainage, subsurface agricultural drainage consisting of larger stones overlaid with gravel which was in turn overlaid with sand; and agricultural drainage channels. The urban drainage system included drainage management of unused domestic water; runoff from thatched roof structures and plaza areas drained to grass or soil covered areas, urban drainage channels combined with stairways, walkways and temple interiors, 129 drain outlets placed in stone retaining walls and building walls, and plaza subsurface strata constructed with rock chips and stones to allow the infiltration of runoff.

The area around Cuzco contains numerous historical sites with examples of water technologies (Fig. 9.15). Pachacuti Inca Yupanqui and his sons designed and implemented water irrigation channels around Cuzco including fountains and culverts for springs. He built Tambo Machay, a spring shrine type structure which according to Spanish historian Bernabé Cobo served as a lodging for Pachacuti when he went hunting (Hemming, 1982). Both Tambo Machay and nearby Puca Pucara are "huacas" ("sacred things" or "anything in nature that is out of the ordinary"), that served primarily for religious services and worships (Bauer, 1998). Bauer also estimated that of the 328 huacas presented in Bernabe Cobo's report 96 (29%) are springs or sources of water and approximately 95 (29%) are standing stones. Other categories included hills and mountain passes, palaces of royalty, tombs, ravines, caves, quarries, stone seats, sunset markers, trees and roads. Tipon, Pisac, Tambo Machay and Puca Pucara are described below in more details.

Tipon is located 23 km southeast of Cuzco at an elevation of 3,560 m above mean sea level. It is made up of 12 terraces that are flanked by polished stone walls. In Tipon water was diverted from the Rio Pukara for irrigation and domestic use. Three irrigation canals (aqueducts) diverted water upstream of Pukara, approximately 1.35 km north of Tipon's central terraces. The main aqueduct diverted water from the river at an elevation of 3,690 m. Slopes of the canal ranged from very steep (30%) to rather flat slopes (2%) of the canal (Wright, 2006). Figures 9.17, 9.18, and 9.19 show portions of the main aqueduct.

Fig. 9.17 Main aqueduct for Tipon. Note the stair steps on the steep terrain (photo copyright by Yuri Gorokhovich). Color version available in Appendix

Fig. 9.18 Elevated portion of the main aqueduct in Tipon (photo copyright by Yuri Gorokhovich). Color version available in Appendix

Fig. 9.19 The source point of the main aqueduct above Tipon (photo copyright by Yuri Gorokhovich). Color version available in Appendix

Figure 9.20 shows the restored Inca principal or main fountain in Tipon, that received water from the main spring in the background. This fountain provided domestic water supply for the noble residents prior to use of the water for any other purpose. Another fountain which supplied water for ceremonial purposes and

Fig. 9.20 Restored Inca fountain in Tipon. This principal fountain receives water from the main spring in the background (photo copyright by Yuri Gorokhovich). Color version available in Appendix

domestic use was located on the side of one of the terraces. This fountain received water from a canal through an approach channel and conduit, and then dropped it into a stone basin. The unused water flowed into another downstream canal. A fountain also was built in the Ceremonial Plaza. Some scholars have concluded that this plaza was a ceremonial reservoir while others believe that the structure could not store water because the walls were not water tight (Wright, 2006). Remains of a canal have been found in the Ceremonial Plaza. Figure 9.21 shows the spring located at the head of the central terrace ravine. Water from the spring was distributed to ceremonial fountains and to irrigated agriculture through canals. Figure 9.22 shows the hydraulic drop structures inset in the terrace walls near the middle of the central terraces.

Pisac is one of the closest major ruins to Cuzco located in Yucay Valley near River Vilcanota (Figs. 9.15 and 9.23) at elevation of 3,400 m. Its history and origin are almost absent for a chronicles, except from few notes by Sarmiento de Gamboa in his description of the valley of Pisac (Hemming, 1982). It is possible that this was a place visited by Pachacuti Inca Yupanqui for pleasure but no historical documents confirm this. Some architectural evidences, such as gates, defensive walls and stone door pegs suggest a possible military role of Pisac. The site structures are scattered along the mountain side and consist of granaries, living quarters, fortified barracks and series of beautiful agricultural terraces with irrigation channels (Fig. 9.24).

In Pisac water was used mainly for irrigation and religious services. The central inti-huatana (The Temple of Sun) in Pisac is surrounded by walls and has a few baths and a water channel (Figs. 9.25 and 9.26). The water channel comes out of the western side of the mountain and is about 20-25 cm wide. It is hard to reconstruct the original water works structure because extensive restoration projects at

Fig. 9.21 Tipon spring located at the head of the central terrace ravine. Water was distributed to ceremonial fountains and to irrigated agriculture through canals (photo copyright by Yuri Gorokhovich). Color version available in Appendix

Pisac "mask" ancient remnants with more pleasing appearance catered for tourists (Fig. 9.27).

Tambo Machay ("resting place" or "resting cave") and Puca Pucara ("red fortress") are located within the walking distance from Cuzco (Fig. 9.15) and within 500 m from each other, separated by the main road from Cuzco to Pisac. Comparing to Machu Picchu, Pisac and Tipon they represent water use technologies on a much smaller scale. Both Tambo Machay (Fig. 9.28) and Puca Pucara (Fig. 9.29) have stone fountains with dual channels. Usually described as fountains, these structures are springs flowing through the calvert build in tightly masoned wall. However, the fountain next to Puca Pucara is dry (Fig. 9.30). Its location on high elevation outside of the Puca Pucara complex suggests that at certain times there was enough groundwater gradient to produce water discharge sufficient for religious or other uses.

One of the important questions regarding water uses a described locations (Machu Picchu, Pisac, Tipon, Tambo Machay, Puca Pucara) is water sources. There are very little or no reliable chronicles available to document these sites from historical perspective and no documentation is available that would describe their water

Fig. 9.22 Hydraulic drop structures near the middle of the central terraces in Tipon (photo copyright by Yuri Gorokhovich). Color version available in Appendix

uses, sources and construction techniques. Fairley (2003) refers to "geologic water storages" for several sites, including Tambo Machay. In hydrology, they are traditionally known as aquifers (for the underground water storage) and watersheds (for the surface waters). Building retaining walls and culverts helped Incas to retain water and reduce discharge. The same technique is widely used around the world now and was already used in ancient times.

Unfortunately, no hydrologic and hydrogeologic studies (including ones with geophysical or drilling techniques) are available to pinpoint the exact source of the water and its geologic and hydrogeologic conditions. Moreover, many sites experienced reconstruction and modification of the original environments and settings. Some current water flows (e.g. in Pisac) are the result of modifications that have nothing to do with, original functionality of the water system. Alternatively, these current flows may represent some ancient water system, yet unknown to us. This uncertainty provides a wide area for scientific hypotheses and discussions. One of the hypotheses relates to the so-called perched aquifers that were used to supply water in various areas around the world, including earlier and late Bronze Age cities in the Mediterranean region (see Chapter 10).

Fig. 9.23 View from Pisac on Vilcanote River and the Yucay Valley (photo copyright by Yuri Gorokhovich). Color version available in Appendix

Fig. 9.24 Terraces in Pisac. There are living quarters located above the terraces (photo copyright by Yuri Gorokhovich). Color version available in Appendix

Fig. 9.25 Central Inti-Huatana in Pisac bordered by the water channel connected to ritual baths (photo copyright by Yuri Gorokhovich). Color version available in Appendix

Fig. 9.26 The segment of restored channel bordering Inti-Huatana in Pisac (photo copyright by Yuri Gorokhovich). Color version available in Appendix

Fig. 9.27 Reconstruction of ancient water works often "masks" the original structure (photo copyright by Yuri Gorokhovich)

Fig. 9.28 Tambo Machay complex consisting of a retaining wall and a calvert with dual channels (*in the center of the picture*) (photo copyright by Yuri Gorokhovich). Color version available in Appendix

Fig. 9.29 Puca Pucara complex (photo copyright by Yuri Gorokhovich). Color version available in Appendix

Fig. 9.30 Dry fountain with dual channels next to Puca Pucara complex. Tambo Machay is located at the upper left corner of the picture (photo copyright by Yuri Gorokhovich). Color version available in Appendix

References

Ackerly N, Howard J, McGuire RH (1987) La Ciudad canals: A study of Hohokam irrigation systems at the community level. Arizona State University, Anthropological Field Studies 17, Department of Anthropology, Arizona State University, Tempe, AZ

Andrews JP, Bostwick TW (2000) Desert farmers at the river's edge: The Hohokam and Pueblo Grande, 2nd ed. Pueblo Grande Museum and Archaeological Park, City of Phoenix, Phoenix, AZ

Bauer B (1998) The sacred landscape of the Inca: The Cuzco Ceque system. University of Texas Press, Austin, TX

Bauer B (1992) The development of the Inca state. The University of Texas Press, Austin, TX
Bribiesca-Castrejon JL (1958) El agua potable en la Republica Mexicana: Los abastecimientos en la epoca prehispanica, Ingeniera Hidraulica en Mexico, 12.2:69–82
Brundage BC (1979) The fifth sun: Aztec gods, Aztec World, Austin and London, The University of Texas Press, Austin, Texas
Coe MM (1993) The Maya, 5th edn. Thames and Hudson, Inc., New York
Coe MM (1994) Mexico: From the Olmecs to the Aztecs, 4th edn. Thames and Hudson, Inc., New York
Crown PL, Judge WJ (eds) (1991) Chaco & Hohokam: Prehistoric regional systems in the American southwest, School of American Research Press, Santa Fe, NM
D'Altroy T (2003) The Incas. Blackwell Publishing
Doolittle WE (1990) Canal irrigation in prehistoric Mexico: The sequence of technological change, University of Texas Press, Austin, TX
Fairley JP (2003) Geologic water storage in pre Columbian Peru. Latin American antiquity 14(2):193–206
Haviland WA (1970) Tikal, Guatemala and Mesoamerican urbanism. World Archaeol 2(2)
Haury EW (1978) The Hohokam: Desert farmers and craftsmen. University of Arizona Press, Tucson, AZ
Hemming J (1982) Monuments of the Incas, Little, Brown, Boston
Hunt RC, Guillet D, Abbott DR, Bayman J, Fish P, Fish S, Kintigh K, Neely JA (2005) Plausible ethnographic analogies for the social organization of Hohokam canal irrigation. American Antiquity 70(3):433–456
Hyslop J (1990) Inka settlement planning. University of Texas Press, Austin, TX
Lister RH, Lister FC (1983) Those who came before. Southwestern Parks and Monuments Association, University of Arizona Press, Tuscon, AZ
Masse B (1981) Prehistoric irrigation systems in the Salt River Valley, Arizona. Science 214(3):408–415
Masse B (1991) The quest for subsistence, sufficiency and civilization in the Sonora Desert. In: Crown PL, Judge WJ (eds) Chaco & Hohokam: Prehistoric regional systems in the American southwest, School of American Research Press, Santa Fe, New Mexico
Matheny RT (1976) Maya lowland hydraulic systems, Science, 193(4254), August 10, 1976
Matheny RT et al. (1983) Investigations at Edzna: Campeche Mexico. New World Archaeological Foundations, Brigham Young University, Provo, UT
Mays LW (2007a) Water resources sustainability. McGraw-Hill, New York
Mays LW (2007b) Water sustainability of ancient civilizations in Mesoamerica and the American southwest. Water Sci Technol: Water Supply, IWA Publishers, 7(1):229–236
Millon R (1983) The place where time began: an archaeologist's interpretation of what happened in Teotihuacan history. In: Berrin K, Pasztory E (eds) Teotihuacan: art from the city of the gods. The Fine Arts Museums of San Francisco, San Francisco
Parlerm A (1955) The agricultural basis of urban civilization in Mesoamerica. In: Steward J H et al.,(eds) Irrigation civilizations. A comparative study. Pan American Union, Social Science Monographs, Washington, DC, pp. 28–42.
Sanders WT (1981) Ecological adaptation in the basin of Mexico: 23,000 B.C. to the present. In: Riefler Bricker V, Sabloff JA (eds) Supplement to the handbook of middle American Indians, Vol. I, Archaeology. University of Texas Press, Austin, pp. 147–197
Sanders WT, Price BJ (1968) Mesoamerica: The evolution of a civilization. New York
Turney O (1929) Prehistoric irrigation in Arizona. Arizona Historical Rev 2(5), Phoenix, AZ
Vivian RG (1974) Conservation and diversion: Water control systems in the Anasazi southwest. In: Downing T, Gibson M (eds) Irrigation impact on society, Anthropological paper No. 25. University of Arizona, Tucson, AZ, pp. 95–112
Vivian RG (1990) The Chacoan prehistory of the San Juan basin. Academic Press, San Diego, CA
Woodbury RB (1960) The Hohokam canals at Pueblo Grande, Arizona, American Antiquity, 26(2)
Woolf G (General Editor) (2005) Ancient civilizations. Duncan Baird Publishers, London

Wright KR, Valencia Zegarra A (1999) Ancient Machu Picchu drainage engineering. J Irrigation Drainage, ASCE, 125(6):360–369
Wright KR (2006) Tipon: Water engineering masterpiece of the Inca Empire. ASCE Press, Reston, Virginia
Wright KR, Valencia Zegarra A (2000) Machu Picchu: A civil engineering marvel. ASCE Press, Reston, Virginia
Wright KR, Kelly JM, Valencia Zegarra A (1997) Machu Picchu: Ancient hydraulic engineering. J Hydraulic Eng, ASCE, 123(10):838–843

Chapter 10
Ground Water Resources and Earthquake Hazards: Ancient and Modern Perspectives

Yuri Gorokhovich and Lee Ullmann

10.1 Introduction

Hydrologic responses to earthquakes such as water level oscillations in monitoring wells and flow changes in streams have been known for decades. However, damage to aquifers and changes in groundwater supplies represents an earthquake hazard that has received relatively little attention from the scientific community. Yet, its impact is high as it leaves well infrastructure without water, results in water pollution, and creates a threat to public health in the aftermath of earthquakes. Although there are relatively few documented cases of the effects of seismic activity on aquifers and groundwater in both the ancient written record and in archaeological studies dealing with the consequences of earthquakes, the use of modern seismic examples and technology could allow for new interpretations of ancient disasters. This chapter reviews known documented modern cases of groundwater supply damages and attempts to analyze similar ancient occurrences in Crete during the Bronze Age.

In 507 A.D. Joshua the Stylite, who composed in the Syriac language, recorded the conditions of the aftermath of the earthquake of 498–499 A.D.:

> In the month of Îlûl (September) there was a violent earthquake, and a great sound was heard from heaven over the land, so that the earth trembled from its foundations at the sound.... and, as some said, a marvelous sign was seen in the river Euphrates and at the hotspring of Abarnệ, in that the water which flowed from their fountains was dried up this day. It does not appear to me that this is false, because, whenever the earth is rent by earthquakes, it happens that the running waters in those places that are cleft are restrained from flowing, and are at times even turned into another direction (Wright, 1882).

This excerpt refers to the earthquake that destroyed the Nikopolis settlement located in Commagene near Euphratensis and is amongst the earliest documented evidences in the Near East for the effect on water supply by earthquakes.

Y. Gorokhovich (✉)
Department of Environmental, Geographical and Geological Sciences, Lehman College, Cuny, Bronx, NY, 10468, USA
e-mail: yuri.gorokhovich@lehman.cuny.edu

L.W. Mays (ed.), *Ancient Water Technologies*, DOI 10.1007/978-90-481-8632-7_10,

The text reports typical earthquake damage patterns, which today are analyzed in the discipline of seismology. Seismology is not only concerned with the study of earthquakes, but also the fluctuations of water level in wells and streams in relation to stresses and strains caused by seismic waves propagating through the rock media, including aquifers (Brodsky et al., 2003; Manga et al., 2003; Matsumoto et al., 2003; Quilty et al., 1995; Roeloffs et al., 1995; Rojstaczer and Wolf, 1994; Rojstaczer and Wolf, 1992). Earthquakes have a variety of hydrological effects including impacts on aquifers, changes in the water table, and alterations in groundwater supply. Seismic events can damage aquifers feeding domestic wells and can interrupt or completely stop their water supply. This affects the population in the post-disaster period when the absence of adequate water supply or the use of well water that was contaminated impacts public health. Earthquake hazards to well infrastructure are common in areas with both low and high levels of seismic risk.

The response to the loss of water supply today in developed and developing countries is similar: additional water resources are brought to affected population by means of transportation until the water supply is repaired. This happened during such events as the Pymatuning earthquake that hit Pennsylvania in 1998 and also the last earthquake in Peru when three coastal cities Ica, Pisco, and Chincha were hit by an earthquake in August 2007. Water for these Peruvian cities had been supplied by tankers (OCHA, 2007).

Though two different countries, the help and remediation was the same: drinking water was supplied from outside sources in various containers and the water supply system was subsequently fixed using local and outside help. This is characteristic for the modern world with fast information flow and technical equipment (drilling machines, motorized transport, generators, etc.) However, besides being examples of human vigor and technological advancement, modern cases of water supply loss due to earthquakes are also a tool that can be used to study the past.

Analysis of current hydrologic disasters caused by earthquakes provides insight into ancient times when the ability to prepare for and respond to natural disasters was limited at best. In the Bronze Age problems with water, such as loss of water sources or change in groundwater supplies, were most probably not interpreted as solely a matter of technical solution, but rather as an omen or a sign from the gods that would have had religious and political repercussions (Driessen, 2001). In antiquity, and more specifically in the Mediterranean World of the Bronze Age, there were three significant types of water sources: springs, wells, and rivers (Zertal, 1988).

Because of the hot and semiarid Mediterranean climate with significant rain falling only in the winter months, between November and February (Crouch, 1993), fluctuations or loss of water or flow output could have had devastating consequences on the cultures that relied on these sources of water. The notion of transporting vital resources such as water and grain in ancient times seems to be somewhat unfeasible; Marc Van de Mieroop explains that even in the late Roman Empire, when an excellent road system covered much of Anatolia, a city could suffer a terrible famine when grain was readily available less than 100 kilometers away (2007). Yet, archaeologists often examine only the structural damage and abandonment of ancient buildings and sites due to earthquakes (Stiros, 1996) and overlook the effects on water resources, which can have far reaching ramifications.

This study attempts to show how modern perspective can provide possible explanations of ancient events. It draws on examples of natural disasters from the 20th century associated with the loss of water due to earthquakes and attempts to superimpose geological interpretation with historical and archaeological evidence from similar incidents of the past. Specifically, we target the ancient Minoans of Crete to illustrate yet another facet that might have prompted both a shift in their cult practices and the decline of the Minoan culture.

10.2 Modern Perspective

10.2.1 Examples of Water Supply Loss Due to Earthquakes

Areas with low seismic activity such as the eastern United States or eastern Canada provide numerous examples of water loss in the local wells as a direct result of seismic activity. One example, the Cornwall Massena earthquake (centered between Massena, New York and Cornwall, Ontario) of 1944 had a magnitude of 5.8 (on the Richter magnitude scale) and had clear hydrological effects on the area. For instance, a large number of wells were left dry in St. Lawrence County causing acute hardship; however, the number of wells and the size of the affected population were not reported (Stover and Coffman, 1993).

In 2002, a 5.0 magnitude earthquake in Essex County, New York resulted in two counties being declared in a state of emergency due to the significant amount of damage that occurred as a direct result from the earthquake, including loss and contamination of well water. The New York State Department of Health was called on to intervene in the aftermath and to assist local governments to ensure the safety of drinking water supplies in both Clinton and Essex counties. However, the number of privately owned wells affected by the earthquake was not reported (Office of the Governor Press Release, 2002).

The most dramatic example of water loss in the area with low seismic risk is the Pymatuning Earthquake of 1998 in northwestern Pennsylvania. The earthquake had a magnitude of 5.2 and resulted in the disruption of more than 120 household supply wells in an area of approximately 50 miles^2 (129.5 sq.km) (Fleeger et al., 1999). In the report published by the United States Geological Survey (USGS) an official observed that 'the most important result of the earthquake was water well damage.' In one well, the water level was documented to have decreased by 40 ft (12.2 m) overnight; this was followed by a sustained period of decline over approximately two months where the level dropped by an additional 55 ft (16.7 m).

Furthermore, water level decline by as much as 100 ft (30.48 m) was reported and the affected wells never gained original water levels. As a result, more than 120 homes were left without water after the earthquake and wells had to be subsequently re-drilled. It took more than one month to drill new wells which were to provide the communities of Greenville and Jamestown with water. In the case of homes that lost well water, most homeowner's insurance policies did not cover water loss, so residents had to drill new wells at their own expense and the process took more than three months (Armbruster et al., 1998).

Apart from earthquakes described in New York, Pennsylvania, and Maine, other states in low seismic risk areas reported similar incidents of wells becoming dry as a result of earthquakes. These included the states of Idaho, Montana, Ohio, and Georgia (USGS, 2003, USGS, 2006; Ohio Department of Natural Resources, 1986; Whitehead et al., 1984). In all of these cases, the assessment of earthquake impact on wells was insufficient, mainly because seismic and hydrogeologic studies were usually conducted separately.

Areas with high seismic activity provide many more examples of water supply disruption by earthquakes. California is a high-seismic activity region and it is one of the most affected states. One of the earliest earthquakes mentioned in relation to the loss of well water in California was the Fort Tejon earthquake on January 5, 1857, which registered a 7.9 magnitude. Hydrological alterations as a result of the Fort Tejon earthquake occurred throughout the region, and changes in well water flow were reported as far north as the Santa Clara Valley (Yewell, 2007).

The 1906 San Francisco earthquake had hydrologic effects similar to those of 1989 Loma Prieta earthquake, including the drying up of many wells in the Pajaro River area of Santa Cruz County (USGS, 1994). The 1989 Loma Prieta earthquake in northern California, magnitude 7.1, left dry wells for residents of San Mateo and Monterey counties (Municipal Services Review for the North County Area of Monterey County, 2004; The South Skyline Association, 2003).

Most recently, in 2003 the San Simeon earthquake (magnitude 6.5) affected California's central coast and resulted in the drying and collapsing of wells, among many other hydrological changes in the region. One of the severely affected was San Luis Obispo County, where cattle is one of the top two commodities produced. In the aftermath of the earthquake, ranchers were forced to trim their herds because of uncertainty in water supply (California Farm Bureau Federation, 2004).

Apart from the incidents described above, where earthquakes have resulted in the drying of wells, changes in well water levels and water quality occur on a widespread basis; they can be associated with earthquakes whose epicenters are located at various distances from wells. For example, multiple earthquakes in Alaska, as well as the earthquake that generated the 2004 Asian tsunami, caused well water fluctuations in the continental United States (Rozell, 2004; USGS, 2005). Changes in water levels in a USGS observation well in Virginia were recorded in association with approximately 50 earthquakes worldwide (usually greater than 6.0 in magnitude) within a 16 month period between 2005 and 2007 (USGS, 2007). In addition to changes in water levels, increased turbidity in wells is also associated with earthquakes and is important because water can be rendered unsafe for drinking.

Unfortunately, lack of data on the location of wells in relation to the topography and the wells' water conditions (i.e. 'dry' wells or 'wet' wells) prevents more substantial research on the effect of earthquakes on wells. The described cases of water supply loss in the US (except the Pymatuning earthquake case) contained no data on wells' water conditions (i.e. 'dry' vs. 'wet') or the location of wells with respect to topography. Moreover, only USGS wells that have monitoring functionality are mentioned in most studies published by hydrogeologists dealing with earthquake effects on groundwater (e.g. Montgomery and Manga, 2003; Roeloffs et al., 1995; Rojstaczer and Wolf, 1992; Rojstaczer and Wolf, 1994). No household wells that lost water due to earthquake activity in the US have been mapped or assessed in a comprehensive way by the specialists.

10.2.2 Geological Background

All above mentioned cases of water loss due to earthquake activity are known from published sources such as newspapers, internet web sites, and scholarly papers/reports. However, detailed geographic data are required to study the effect of earthquakes on hydrologic environments. Such data should include the geographic location of wells, their characteristics regarding water level before and after an earthquake, impact on households, condition of casing (e.g. collapsed vs. intact), and detailed description of water quality. While there are many documented studies of earthquake effects on water level in USGS monitoring wells, there are only a few cases known to the scholarly community that are well documented with maps and data on the conditions of the domestic wells serving water supply. Both geological and cartographic data are critical to understanding the environmental conditions that create a hazard and consequently lead to a disaster.

Data on water loss in domestic wells and associated data on geology and topography would allow scientists to create models based on geological and environmental conditions to study the repercussions of seismic events in the past and present. For example, we already know that the potential impact of an earthquake on domestic water supply can cause reduction in water discharge in wells located on top of ridges and mountains and increase in discharge in wells located in valleys and low-lying areas. However, we still do not know specifics about hydrogeologic conditions that cause these effects; furthermore, we do not know geological and hydrologic conditions that set the stage for these events.

Some of the influencing geological factors and environmental conditions related to the effect of seismicity on loss of water supply in wells are discussed in Gorokhovich and Fleeger (2007) using the Pymatuning earthquake as an example that was well documented and described previously by Armbruster et al. (1998) and Fleeger et al. (1999). The model which we will be referring to later as the 'Pymatuning model' contains two potential hazardous conditions: (1) Heterogeneity of geological material composing aquifers or existence of a perched aquifer; (2) Location of wells on any high elevation (plateau, slopes above valleys, etc.) that can be treated as a "hydrologic island", i.e. a structure that has a limited and geographically narrow source of recharge (Toth, 1963; Norvatov and Popov, 1961).

These two conditions favor a situation when aquifers located above the valley bottom can lose water after an earthquake due to increase in their permeability. This increase can be drastically exacerbated by the heterogeneity of the aquifer material that can cause an unexpectedly rapid increase in permeability. This, in turn, causes fast discharge of groundwater down the valley bottom with the subsequent increase in river discharges or overflow of water in wells located at the valley bottom or in its vicinity.

The Pymatuning model can also explain the effects of the Matsushiro earthquake that took place in Japan in 1965 and produced series of seismic shocks through March 1968, causing among other disasters (such as faults and landslides) a loss of water in two villages (Terakawa and Matsuo, 1996). A published map and GIS topographic data helped restore the geological settings in the area affected by the earthquake. Figure 10.1 shows a recreated map of the affected area between the cities of Nagano and Ueda using data from Terakawa and Matsuo (1996) and global digital elevation SRTM (Shuttle Radar Topographic Mission) model. Oval symbols show areas where water discharge increased and decreased.

According to the geologic map of Japan posted by the Geologic Survey of Japan (GSJ, 2008), area between Ueda and Suzaka (Fig. 10.1) consists mainly of mafic and plutonic volcanic rocks overlaying early miocene marine and non-marine sedimentary rocks. These rocks are represented by porphyrites, diorites, pebbly mudstone, and sandstone covered by alluvial deposits; high quantities of clays and clay minerals were found on slopes within the Matsushiro area. These sediments exacerbated slide-prone conditions (Morimoto et al., 1967). Intrusions of diorite-porphyrite penetrated sedimentary rocks during the Miocene and later during Pliocene and Pleistocene were covered by Pleistocene volcanic complexes (Tsuneishi and Nakamura, 1970). Accumulation of sediments still continues in the area and results in a composition of 328–656 ft (100–200 m) thick alluvial fans and unconsolidated sediments (Ono, 1967; Nakamura and Tsuneishi, 1967).

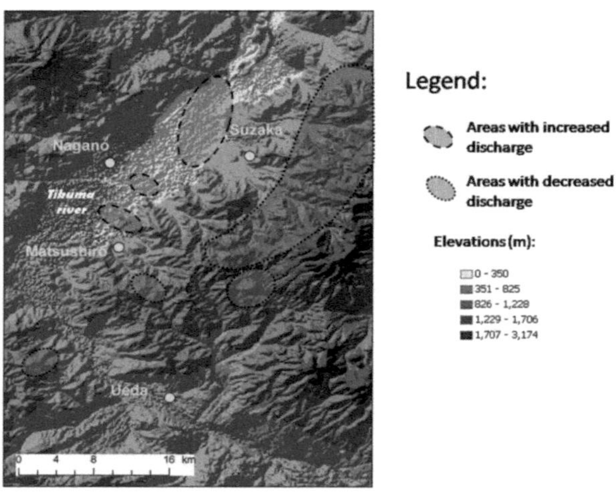

Fig. 10.1 Geographic settings of the Matsushiro earthquake (areas of water discharge/overflow were adapted from Terakawa and Matsuo, 1996)

Geological conditions in the Matsushiro area provide favorable environment for heterogeneous aquifers with variable permeability. In such conditions, aquifers could easily loose water during earthquakes. The terrain is mountainous and some small mountains with elevations of less than 3,280 ft (1,000 m) are separated from more sizeable mountains with much higher elevations (more than 9,842 ft or 3,000 m a.s.l.) located in the east.

Figure 10.1 shows that areas with dried up wells are on high elevations (2,625–3,280 ft or 800–1,000 m a.s.l.) while areas with excessive flows are in the valley near the Tikuma river. Both conditions (high elevations and heterogeneous aquifers) are in accordance with the Pymatuning model, which was also applied to the interpretation of possible causes of crisis on Crete (Gorokhovich and Fleeger, 2007).

10.3 Ancient Perspective

The destruction of the Bronze Age Minoan culture of Crete is heavily debated, with theories spanning the gamut from invaders to tsunamis; yet, all these theories have not brought us any closer to understanding what led to the demise of this fascinating culture. Following the work of Jan Driessen and Colin F. Macdonald (1997), it is clear that an array of factors contributed to the downfall of the Minoans. In particular, 'the archaeological evidence suggests a severe economic dislocation,' triggered by an increase in seismic activity during the Minoan period (Monaco and Tortorici, 2004; Nur and Cline, 2000). However, few if any of the more recent studies that analyzed factors that led to the demise of the Minoans have examined possible effects that earthquakes may have had on groundwater and water supplies during the Late Minoan IA-IB period (Gorokhovich and Fleeger, 2007; Gorokhovich, 2005). Our theory argues that collapse of the Late Minoan IA-IB well infrastructure in main palatial settlements, such as Phaistos and Knossos, was the major factor that precipitated socio-economic changes at these centers and might have played a role in the formation of what Jan Driessen has referred to as 'crisis cults' as a method to cope with the change in water supply (Driessen, 2001).

Our theory is based on similarities between the geological settings of the modern day earthquakes at Pymatuning (Pennsylvania) and Japan and those of ancient Crete. All geographic locations share geological heterogeneity of aquifer material that could provide favorable conditions for appearance of so-called 'perched aquifers.' Perched aquifers can provide water on hilltops, unlike regular deep or shallow aquifers that discharge directly into the river valley. The evidence from the previous section Modern Perspective shows that earthquake activity can trigger structural changes in geologic media and destroy the ability of perched aquifers to feed local wells.

The significance of water and springs for rituals and cults is readily apparent when studying cultures of the ancient world, yet due to absence of written documents in Crete it is often difficult to provide evidence for the Minoans' affinity for water (Rutkowski, 1986). However, Alan Peatfield, an Aegean archaeologist, has been able to illustrate the importance of liquid and its ritual manipulation as a structural element in Minoan religion through the use of images and the placement of Minoan ritual sites (1995). As he explains, 'close to many Minoan peak sanctuaries there are wells and springs, with sherds scattered around, e.g. Jouktas, Kophinas, and Karphi' (1995). Moreover, at the sanctuary of Kato Syme, on the southern slopes of Mt. Dikte, an abundant spring cuts through the part of the shrine and must have influenced the choice for the establishment of the cult in this remote location (Lebessi and Muhly, 1990; Peatfield, 1995).

The connection between Minoan shrines and springs suggests use of water as part of the Minoan belief system. The Minoans' emphasis on the natural world is not surprising considering that they were predominantly an agrarian society that depended on rain, rivers and springs to provide for their sustenance. Peatfield continues by describing how Minoan society functioned: 'in a sense, therefore, fertility underpins the whole social order, and ensures its continuity. Fertility is also fundamental to social hierarchy, in that access to and control of its products provide the economic base of wealth and prestige... Therefore, any threat to that fertility is also a metaphorical threat to society – a crisis.' (1995) These statements provide direct support to our theory that the loss of spring and ground water as a result of seismic activity could create a crisis situation that would reverberate through both religious and political spheres of Minoan society.

It is hard to gain a clear perspective on storage and use of water installations in Crete during the Late Minoan (LM) period for numerous reasons. For instance, there is a great debate over the function and dating of *koulouras* (the Greek word for something round and hollow) or the wide and/or deep round structures found at sites such as Knossos, Mallia, Myrtos-Pyrgos, and Phaistos ranging from the Early Minoan through the Late Minoan (LM) period, which would seem to be the repositories for collecting water (Fig. 10.2). Some argue that they were cisterns (Watrous et al., 2004; Cadogan, 2007), others granaries (Strasser, 1997 [with regard to Mallia only]; Halstead, 1997; MacGillivray, 1994), and Evans suggested that at Knossos they were rubbish pits (1935).

Fig. 10.2 *Koulouras* in Phaistos palace

One should also consider the possibility that over time these structures may have had more than one function. There is some consensus on two LM I plastered round structures with paved bottoms, which allowed the ground water to seep in between the stones, from Archanes-Tourkoyeitonia and in the east wing of the Palace at Zakro that functioned as a well in antiquity (Cadogan, 2007). Thus, getting a better understanding of the effects of geological hazards in ancient times requires a joint, multidisciplinary effort from hydrogeologists, archaeologists, geologists, and historians to investigate geological and archaeological field data, analyze contemporary events in nearby locations, and make comparisons with other geographic places in order to elicit a solution. Unfortunately, most studies to date are conducted by scholars in a single field and do not utilize unique tools, approaches, and theories that a multidisciplinary work can offer.

That being said, a geological study of Phaistos locality by Watrous et al. (1993) revealed that the hill supporting the Palace of Phaistos consists of mixed marl and conglomerate deposits from Upper Miocene. By geological definition, marl is a sedimentary rock consisting of clay and calcium carbonate, usually formed by diagenesis from marine deposits. The heterogeneity of this material favors existence of perched aquifers, which would provide water for domestic needs. Findings of wells and distributed water mains in Phaistos serve as evidence to the existence of an ancient water supply system. The river Ieropotamos at the foothill of Phaistos provided additional water to the palace; however, today both the river and the wells are dry.

It is interesting to note that later on in history, in the time of Justinian (527–565 A.D.), earthquakes were still affecting life of the inhabitants of Crete. At Gortyn, the capital of Roman and Early Byzantine Crete, two inscriptions, I.C. 461 and I.C. 465 (Codex Vaticanus Graecus, 1759), found in situ, both dated to the second half of the 6th century, acknowledge a certain Georgios who returned the water supply to the town (Di Vita, 1996). These two inscriptions lead one to believe that the extensive damage of the old underground water supply system and its eventual replacement by a new aqueduct were the results of a seismic disaster in approximately 560 A.D. This suggests that loss of water in the settlements of Crete in ancient times was a recurrent event throughout the history of the island.

A comparison of Crete with other locations in the Mediterranean basin provides more insight into the ancient perspective by reviewing the link between geologic structure and settlement patterns. In a study of Malta, Louis Cassar (1997) noted an important role of local marl deposits on settlement patterns. He wrote: 'The relatively thin stratum of Blue Clay, in fact a limestone marl, is said to have made early settlement on the islands possible. Since clay is impervious, a large number of springs occur through the percolation of rain water, being significant as a domestic water source and as the basis of a life-giving zone of irrigated farmland. In historic times, artesian wells were dug through the upper layers in order to tap the water trapped above the clay.'

In another study of the city of Priene, D.P. Crouch (1996) mentioned subsurface streams in the upland areas with carbonate rocks. According to the author, these streams provided more than 80% of houses with water tapped by pipelines from the mountain. She also referred to the study of Meulenkamp (Meulenkamp et al., 1972) who pointed out that "bioclastic limestones ... with intercalation and displaced boulders of bluish clays or marls ... are found at more than 200 meters above sea level." Crouch indicated that water from this height was brought down into the city of Rhodes and its acropolis. The acropolis also served as water storage via the use of grottoes.

The available evidence suggests that ancient civilizations used perched aquifers as a water resource. The relative fragility of perched aquifers, their heterogeneity and limited spatial extent can explain both perched aquifers' important role for water supply and cult and their vulnerability to seismic hazards. Although the direct historical evidence is scarce, the loss of water supply induced by earthquake activity could have caused the relocation of people in ancient times, forcing them to find new sources of water, especially if the resilience of the society was weakened by social, economic, political and religious reasons.

10.4 Discussion

Examples from previous section on Modern and Ancient Perspectives make it apparent that damage to hydrological resources due to seismic activity is a clear and present danger. For example, 37% of the United States water consumption from public sources is based on groundwater (Hutson et al., 2005). Fresh ground-water withdrawals in the United States increased by 14% between 1985 and 2000, and an estimated 242 million people nationwide, or about 81% of the US population, depend on water from public suppliers. These figures indicate that use of groundwater for the national public water supply is significant and that earthquake hazards on groundwater sources could have high social and economic impacts in various geographic regions throughout the US. Yet, the effects of earthquakes are relatively undocumented both in the ancient and modern world.

Regarding earthquake hazards on well infrastructure, the proportion of locations affected and associated economic and social consequences remain unknown. A body of research that quantifies the hydrological impact of earthquake hazards has emerged; however, there is still no systematic research on social and economic consequences of these hazards and no synthesis to evaluate risk levels. Moreover, there are no governmental committees or units that can be assigned the task of assessing this specific hazard because of the interdisciplinary nature of factors contributing to the problem. Furthermore, there is a lack of data regarding the issue of recurrent recharge after the loss of water. This would be an extremely important piece of information for both modern and ancient perspectives as it would allow one to examine: (1) Whether or not an aquifer can maintain the same recharge as in the past; (2) What the recurrence interval for the aquifer recharge (e.g., years, centuries) is?

Another area worth investigating is the extent of the effects of an earthquake on water supply are. For instance, what are critical earthquake parameters, such as magnitude and proximity to the earthquake epicenter, that represent maximum risk for wells? This information would be helpful in determining the location of wells at risk and estimating spatial extent of the potential damage caused by an earthquake. For example, it has been argued that in Crete earthquake damage is often localized and thus would not have accounted for damage throughout the island in the LM IB period (Soles, 1999). However, this refers to the damage sustained by the structures above the ground and not to water resources that are spread and located within different topographic and geomorphological units that can have different levels of resilience to the seismic wave effect. The only known studies related to this area of investigation were done first by Dobrovolsky et al. (1979) and later by Montgomery and Manga (2003) who related changes in stream and groundwater flow to earthquake magnitude and proximity. The study conducted by Montgomery and Manga (2003) showed that "the distance to which earthquake-induced changes in water levels in wells have been reported increases systematically with earthquake magnitude". While this is an important conclusion, it only slightly advances the quantitative identification of the earthquake impact on water supply.

Geologic evidence and factors that contribute to the understanding of earthquake effects on groundwater supply come from modern studies alone. The best documented case is the Pymatuning earthquake. It provided geological and geomorphological criteria for understanding the mechanism of groundwater effects caused by the earthquake activity. While the Pymatuning model is the only case that has a comprehensive description and documentation (in addition to the Matsushira earthquake case that is less documented, but at least offers a geographic map of well locations), geologic factors described in Fleeger et al. (1999) provide a basis for generalization of geomorphologic and structural geological factors applicable to interpretation of both modern and ancient cases.

There is a definite need for a multidisciplinary study which will incorporate geological, hydrogeological, social, and economic factors in the existing known cases of well water disruption due to earthquake activity. Efforts must be undertaken to ensure complete reporting on changes in well water levels with seismic events. This practice will help scientific community evaluate risk factors, preparedness measures and mitigation strategies, and develop requirements and protocols for the assessment of the impact of earthquake hazards on well infrastructure. We need to gain a more complete understanding of the social and economic ramifications of hydrological damage caused by earthquakes in our own time in order to obtain a clearer picture of similar events in ancient times.

10.5 Conclusions

The modern perspective on hazards associated with water supply shortage due to earthquakes reveals that modern societies can deal with this problem by: (1) Increasing depth of old wells or re-drilling new wells; (2) Fixing broken water distribution systems within a relatively short time; (3) Bringing water supplies from outside the area of disaster by various means of transportation (from trucks to vessels and aircraft); (4) Using a variety of containers (from water bottles to tankers) to store water; (5) Providing timely information to potential donors and international humanitarian organizations. These measures exclude potential possibilities of a population to move or migrate due to the shortage of water supply since interruption of the supply is only temporary due to the technical mobilization of the post-disaster management efforts, as described above. In an interview regarding the Pymatuning earthquake of 1996, one of the residents was asked: 'What would you do if you did not have supplied bottle water and help from the fire department that provided daily water for bath tubs?' The answer was: 'I would move ...'

Modern cases of disrupted water supply due to earthquake activity provide us with unparalleled availability of scientific data and research results on how earthquake activity affects groundwater supplies. An abundance of reports, papers, and media news offers a wealth of general information, but scientific information is often limited because scientists collect data pertinent only to their specific disciplines. For example, seismologists record seismic activity in attempts to predict future risks due to aftershocks, geologists record changes in geologic structures to improve seismic zonation and safety of infrastructure and settlements, etc. If there are any hydrologic effects of an earthquake, a hydrologist might be involved as well. However, there are only a few reports that combine multidisciplinary findings into one study.

Available, albeit limited, multidisciplinary studies allow us to summarize the following high-risk geological factors related to hydrologic effects of earthquakes: existence of heterogeneous or perched aquifers and location of wells on high elevations above river valleys. Though these findings are derived from only two studies where locations of wells that went dry and wells that overflowed were mapped (i.e. Fleeger et al., 1999; National Committee for Countermeasures Against Landslides, 1969), other reports and studies provide similar evidence. 'The present is the key to the past' (Lyell, 1830): modern geological findings and established models, like the Pymatuning model, can be used for analysis and interpretation of historic events in conjunction with archaeological data to demonstrate potential effects of earthquakes on ancient settlements.

The proportion of locations affected and associated economic and social impact remain unknown; no systematic research exists on the associated risks. This applies to both the ancient and modern perspectives.

References

Armbruster JG, Barton H, Bodin P et al. (1998) Preliminary results from the investigation of the Pymatuning earthquake of September 25, 1998. Pa. Geology 29(4):2–14

Brodsky EE, Roeloffs E, Woodcock D et al. (2003) A mechanism for sustained groundwater pressure changes induced by distant earthquakes. J Geophys Res 108(B8):7–1–7–10

Cadogan C (2007) Water management in Minoan Crete, Greece: the two cisterns of one Middle Bronze Age settlement. Water Sci Technol: Water Supply 7(1):103–111

California Farm Bureau Federation (2004) Central Coast earthquake still challenging ranchers

Cassar L (1997) Settlement patterns in the Maltese Islands: from early colonization to pre-independence. GeoJ 41(2):137–144

Crouch DP (1996) Environmental geology of ancient Greek cities. Environ Geol 27:233–245
Crouch DP (1993) Water management in ancient Greek cities. Oxford University Press, New York
Di Vita A (1996) Earthquakes and civil life at Gortyn (Crete) in the period between Justinian and Constant II (6–7th century A.D.). In: Stiros S, Jones RE (eds) Archaeoseismology. Fitch Laboratory occasional papers, 7. Institute of Geology & Mineral Exploration: British School at Athens, Athens, pp 45–50
Dobrovolsky I, Zubkov S, Miachkin V (1979) Estimation of the size of earthquake preparation zones. Pure Appl Geophys 117(5):1025–1044
Driessen J (2001) Crisis cults on Minoan Crete? In: Laffineur R, Hägg R (eds) *POTNIA*: Deities and religion in the Aegean Bronze Age, Aegaeum 22. Université de Liège, Histoire de l'art et archéologie de la Grèce antique, Liège (Belgium), pp. 361–369
Driessen J, Macdonald C (1997) The troubled island: Minoan Crete before and after the Santorini eruption. Aegaeum 17. Université de Liège, Histoire de l'art et archéologie de la Grèce antique, Liège (Belgium)
Evans A (1935) The Palace of Minos at Knossos, vol. IV, part I. Macmillan and Co., London
Fleeger GM, Goode DJ, Buckwalter TF et al. (1999) Hydrologic effects of the Pymatuning earthquake of September 25, 1998 in northwestern Pennsylvania. USGS Water Resources Investigation Report 99–4170.
Gorokhovich Y (2005) Abandonment of Minoan palaces on Crete in relation to the earthquake induced changes in ground water supply. J Archeological Sci 32(2):217–222
Gorokhovich Y, Fleeger G (2007) Pymatuning earthquake in Pennsylvania and Late Minoan crisis on Crete. Water Sci Technol: Water Supply 7(1):245–251
GSJ (2008) Geologic Map of Japan, 1:200,000. http://riodb02.ibase.aist.go.jp/db084/Hokurikuzoom_e2.html. Cited 7 April 2009
Halstead P (1997) Storage, strategies, and states of prehistoric Crete: A reply to Strasser. J Mediterranean Archaeol 10(1):103–107
Hutson S, Barber N, Kenny J (2005) Estimated use of water in the United States 2000. USGS Circular 1268. http://pubs.usgs.gov/circ/2004/circ1268/. Cited 7 April 2009
Lebessi A, Muhly P (1990) Aspects of Minoan cult sacred enclosures: The evidence from the Syme sanctuary (Crete). Archäologischer Anzeiger 3:315–336
Lyell Charles (1830) Principles of geology (I). John Murray, London
MacGillivray JA (1994) The early history of the palace at Knossos (MMI-II). In: Evely D, Hughes-Brock H, Momigliano N (eds) Knossos: A Labyrinth of History: Papers in Honour of Sinclair Hood. British School at Athens, London, pp. 45–55
Manga M, Brodsky EE, Boone M (2003) Response of streamflow to multiple earthquakes. Geophys Res Lett 30(5):18–1–18–4
Matsumoto N, Kitagawa G, Roeloffs E (2003) Hydrological response to earthquakes in the Haibara well, Central Japan – I Groundwater level changes revealed using state space decomposition of atmospheric pressure, rainfall and tidal responses. Geophys J Int 155(3):885–898
Meulenkamp JE, de Mulder EFJ, Van de Weerd A (1972) Sedimentary history and paleogeography of the Late Cenozoic of the Island of Rhodes. Z Deutsch Geol Ges 123:541–553
Monaco C, Tortorici L (2004) Faulting and effects of earthquakes on Minoan archaeological sites in Crete (Greece). Tectonophysics 382:103–116
Montgomery RD, Manga M (2003) Streamflow and water well responses to earthquakes. Science 300:2047–2049
Morimoto R, Nakamura K, Tsuneishi Y et al. (1967) Landslides in the epicentral area of the Matsushiro earthquake swarm – Their relation to the earthquake fault. Bull Earthquake Res Institute 45:241–263
Municipal Services Review for the North County Area of Monterey County (2004) Prepared by Cypress Environmental and Land Use Planning in consultation with Ifland Engineers, Inc. for LAFCO of Monterey County http://www.co.monterey.ca.us/lafco/N%20County%20MSR%20Sept%202004%20CD%20Version-c.pdf. Cited 7 April 2009

Nakamura K, Tsuneishi Y (1967) Ground cracks at Matsushiro probably of underlying strike-slip fault origin, II–The Matsushiro earthquake fault. Bull Earthq Res Inst 45: 417–471

National Committee for Countermeasures against Land Slides (1969) Matsushiro Earthquake and the Land Slide (in Japanese)

Norvatov AM, Popov OV (1961) Laws of formation of minimum stream flow. Bull Int Assoc Sci Hydrol (6)1:20–27

Nur A, Cline EH (2000) Poseidon's horses: plate tectonics and earthquake storms in the Late Bronze Age Aegean and eastern Mediterranean. J Archaeological Sci 27(1):43–63

Ono Y (1967) Electrical sounding at Matsushiro earthquake district (I). Notes Coop Res Disas Prev 5:23–27 (in Japanese with English abstract)

OCHA (2007) Peru: Earthquake. OCHA-06:21-Aug-07 Situation Report No. 6 http://cidi.org/disaster/07b/ixl66.html. Cited 7 April 2009

Office of the Governor Press Release, New York (2002) http://www.state.ny.us/ governor/press/year02/press02.html. Cited 7 April 2009

Ohio Department of Natural Resources (1986) January 1986 Northeastern Ohio Earthquake. http://www.dnr.state.oh.us/geosurvey/gen/seismic/sum86/tabid/7896/Default.aspx. Cited 7 April 2009

Peatfield A (1995) Water, fertility, and purification in Minoan religion. In: Morris C (ed) Klados: Essays in Honour of J.N. Coldstream. Bulletin of the Institute of Classical Studies Suppl. 63:217–227

Quilty EG, Farra CD, Galloway DL et al. (1995) Hydrologic effects associated with the January 17, 1994 Northridge, California Earthquake. USGS Open file report 95–183

Roeloffs E, Danskin WR, Farar CD et al. (1995) Hydrologic effects associated with the June 28, 1992 Landers, California, earthquake sequence, USGS Open-file report 95–42

Rojstaczer S, Wolf S (1992) Permeability changes associated with large earthquakes: an example from Loma Prieta, California, 10/17/89. Geology 20:211–214

Rojstaczer S, Wolf S (1994) Hydrologic changes associated with the earthquake in the San Lorenzo and Pescadero drainage basins, USGS Report 'The Loma Prieta, California, Earthquake of October 17, 1989: Strong Ground Motion and Ground Failure'

Rozell N (2004) Earthquakes near and far shake up Alaska wells. Alaska Science Forum Article #1804: May 24, 2006

Rutkowski B (1986) The cult places of the Aegean. Yale University Press, New Haven

Soles JS (1999) The collapse of Minoan civilization: the evidence of the broken ashlar. *POLEMOS*: Le contexte guerrier en Égée à l'Âge du Bronze. Aegaeum 19. Université de Liège, Histoire de l'art et archéologie de la Grèce antique, Liège (Belgium)

Stiros SC (1996) Identification of earthquakes from archaeological data: methodology, criteria and limitations. In: Stiros S, Jones RE (eds) Archaeoseismology. Fitch Laboratory occasional papers, 7. Institute of Geology & Mineral Exploration: British School at Athens, Athens, pp. 129–152

Stover CW, Coffman J (1993) Seismicity of the United States, 1568–1989, U.S. Geological Survey Professional Paper 1527. United States Government Printing Office, Washington DC

Strasser T (1997) Storage and states on prehistoric Crete: the function of the Koulouras in the first Minoan palaces. J Mediterranean Archaeol 10(1):73–100

Terakawa A, Matsuo O (1996) Earthquakes. In: Singh VP (ed) Hydrology of disasters, water science and technology Library 24. Kluwer Academic Publishing, Netherlands, pp. 427–433

The South Skyline Association (2003) Skylines Newsletter July, 2003. http://www.southskyline.org/Skylines/2003-07/2003-07.pdf. Cited 7 April 2009

Toth J (1963) A theoretical analysis of groundwater flow in small drainage basins. J Geophys Res 68(16):4795–4812

Tsuneishi Y, Nakamura K (1970) Faulting associated with the Matsushiro swarm earthquakes. Bull Earthquake Res Institute 48:29–51

USGS (1994) The Loma Prieta, California, Earthquake of October 17, 1989-Hydrologic Disturbances. Rojstaczer S (ed) USGS Professional Paper 1551-E. United States Government Printing Office, Washington DC

USGS (2003) More to know about Alabama-Georgia earthquake http://www.usgs.gov/newsroom/article.asp?ID=190. Cited 7 April 2009

USGS (2005) USGS Florida groundwater monitoring wells respond to Indonesian earthquake http://fl.water.usgs.gov/Tampa/earthquakeresponse.html. Cited 7 April 2009

USGS (2006) Earthquake history of Montana http://earthquake.usgs.gov/regional/states/montana/history.php. Cited 7 April 2009

USGS (2007) Hydrologic response in well 27F2 SOW 019 to worldwide earthquakes. USGS Virginia Water Science Center Jan 22, 2007 http://va.water.usgs.gov/earthquakes/index.htm. Cited 7 April 2009

Van de Mieroop M (2007) The eastern Mediterranean in the age of Ramesses II. Blackwell Publishing, Malden, MA

Watrous LV, Hadzi-Vallianou D, Pope K et al. (1993) A survey on Western Mesara Plain in Crete: preliminary report of the 1984, 1986 and 1987 seasons. Hesperia 62(2):191–248

Watrous LV, Hadzi-Villianou D, Blitzer H (2004) The Plain of Phaistos: cycles of social complexity in the Mesara region of Crete. Monumenta Archaeologica 23, Costen Institute of Archaeology, University of California, Los Angeles

Whitehead RL, Harper RW, Sisco HG (1984) Hydrologic changes associated with the October 28, 1983 Idaho earthquake. Pure Appl Geophys 122(2–4):280–293

Wright W (1882) The Chronicle of Joshua the Stylite. University Press, Cambridge

Yewell J (2007) Even Faults Have Virtues. The Pinnacle, Orange County Register, January 5, 2007 http://homepage.mac.com/jyewell/iblog/B740045460/ . Cited 7 April 2009

Zertal A (1988) The water factor in the Israelite settlement process in Canaan. In: Heltzer M, Lipinski E (eds) Society and Economy in the Eastern Mediterranean (1500–1000 BCE), Proceedings of the International Symposium, University of Haifa, Haifa 1987 (Orientalia Loevenansia Analecta 23) Peeters, Leuven, pp. 341–357

Chapter 11
Lessons from the Ancients on Water Resources Sustainability

Larry W. Mays

11.1 Introduction: Today's Water Crisis

At the beginning of this new millennium we have a water crisis which threatens human's existence in many parts of the world. One might ask, how sustainable is it to live in a world where approximately 1.1 billion people lack safe drinking water, approximately 2.6 billion people lack adequate sanitation, and between 2 and 5 million people die annually from water-related diseases (Gleick, 2005)?

UNICEF's Report, The State of the World's Children 2005: Children Under Threat, provides an analysis of seven basic "deprivations" that children feel and that powerfully influence their futures. UNICEF concludes that more than half the children in the developing world are severely deprived of one or more of the necessities essential to childhood: adequate shelter, sanitation, access to safe water, access to information, health care services, school, and food. Of these three are directly related to water:

- 500 million children have no access to sanitation
- 400 million children do not have access to safe water
- 90 million children are severely food-deprived

There are approximately 2.2 billion children (with 1.9 billion living in the developing world and about 1 billion living in poverty) in the world so approximately one in five in the world does not have access to safe water.

The looming present day water crisis must be faced using traditional knowledge and techniques inherited from the past in addition to our present day technological capabilities for more sustainable ways of dealing with water scarcity, particularly in developing parts of the world. Many present day water problems could be solved using the traditional methods developed and used for hundreds of years by the ancients. In parts of the western world, the philosophy of "having it all and all at

L.W. Mays (✉)
School of Sustainable Engineering and the Built Environment, Arizona State University, Tempe, AZ 85287-5306, USA
e-mail: mays@asu.edu

L.W. Mays (ed.), *Ancient Water Technologies*, DOI 10.1007/978-90-481-8632-7_11,

once" unfortunately has spread around the world. This has blinded many people of the forgotten sustainable ways of the ancients. So that in reality highly advanced methods are not required to solve many water problems, particularly in many of the poor and developing parts of the world. A large part of the future will be to live in concert with nature, not trying to defy nature.

11.2 Water Resources Sustainability

> Water resources sustainability is the ability to use water in sufficient quantities and quality from the local to the global scale to meet the needs of humans and ecosystems for the present and the future to sustain life, and to protect humans from the damages brought about by natural and human-caused disasters that affect sustaining life (Mays, 2007).

Because water impacts so many aspects of our existence, there are many facets that must be considered in water resources sustainability including:

- Water resources sustainability includes the *availability of freshwater supplies* throughout periods of climatic change, extended droughts, population growth, and to leave the needed supplies for the future generations.
- Water resources sustainability includes having the *infrastructure*, to provide water supply for human consumption and food security, and to provide protection from water excess such as floods and other natural disasters.
- Water resources sustainability includes having the *infrastructure* for clean water and for treating water after it has been used by humans before being returned to water bodies.
- Water sustainability must have adequate *institutions* to provide the management for both the water supply management and water excess management.
- Water sustainability must be considered on a *local, regional, national, and international basis*.
- To achieve water resources sustainability the principles of *integrated water resources management (IWRM)* must be implemented.

Sustainable water use is "the use of water that supports the ability of human society to endure and flourish into the indefinite future without undermining the integrity of the hydrological cycle or the ecological systems that depend on it" (Gleick et al., 1995). The following seven sustainability requirements were presented:

- A basic water requirement will be guaranteed to all humans to maintain human health.
- A basic water requirement will be guaranteed to restore and maintain the health of ecosystems.
- Water quality will be maintained to meet certain minimum standards. These standards will vary depending on location and how the water is to be used.
- Human actions will not impair the long-term renew ability of freshwater stocks and flows.

- Data on water-resources availability, use and quality will be collected and made accessible to all parties.
- Institutional mechanisms will be set up to prevent and resolve conflicts over water.
- Water planning and decision making will be democratic, ensuring representation of all affected parties and fostering direct participation of affected interests.

Many of the ancients practiced sustainable water use through building water structures that were adapted to the environment and were fitted into nature. One excellent example was the use of qanats. Qanats rely upon distributing water which is recharge to the groundwater from precipitation. Today we have lost this sense of sustainable water use. By the drilling of wells and pumping beyond the recharge capabilities we have exhausted many aquifers that can no longer provide water supplies. This has even occurred in parts of the world that have qanats, drying them up so they are no longer useable.

11.3 Ancient Water Conflicts

11.3.1 Conflicts Over Water in Ancient History

Water has been a tool of war, a tool of conflict, and a target of conflict. Gleick (2005) presented a chronology of conflict over water in the legends, myths, and history of the ancient Middle East. Biblical accounts (Exodus) include Moses (ca. 1200 B.C.) damming a tributary of the Nile to prevent the Egyptians from reaching the Jews as they retreated through the Sinai. Another is when Moses and retreating Jews were trapped between Pharaoh's army and the Red Sea. Moses parts the waters of the Red Sea so the retreating Jews can escape with the waters closing behind them to cut off the Egyptians.

11.3.2 Ancient Jerusalem's Water Supply Systems

One of the key problems in developing cities in Judah and Israel during the Iron Age was to ensure a continuous water supply in both peacetime and in time of war. In the City of David (Ancient Jerusalem) the only perennial source of water was the Gihon spring in a cave located in the Kidron Valley (Fig. 11.1a), which was beyond the fortified city (Cahill and Tarler, 1994). This spring does not maintain a constant flow, but has intermittent flows (varying with season of the year) through cracks in the cave floor. In antiquity three subterranean water supply systems were developed to obtain the water and store, distribute, and protect the water (Cahill and Tarler, 1994). The strategic water supply systems were Warren's shaft, the Siloam channel, and Hezekiah's tunnel, shown in Fig. 11.1.

Warren's shaft (see Fig. 11.1b), named after the discoverer Ch. Warren, was the earliest of the three strategic water supply systems (10th or 9th centuries B.C.) (Shiloh, 1994). This system allowed residents of the city to obtain water from the

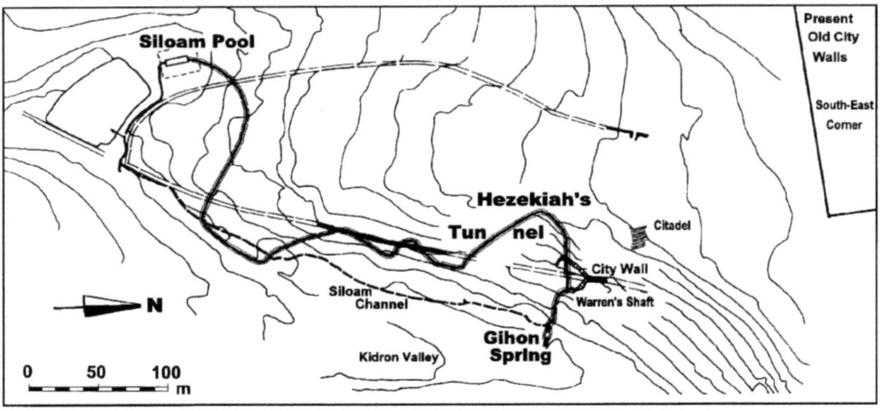

(a)

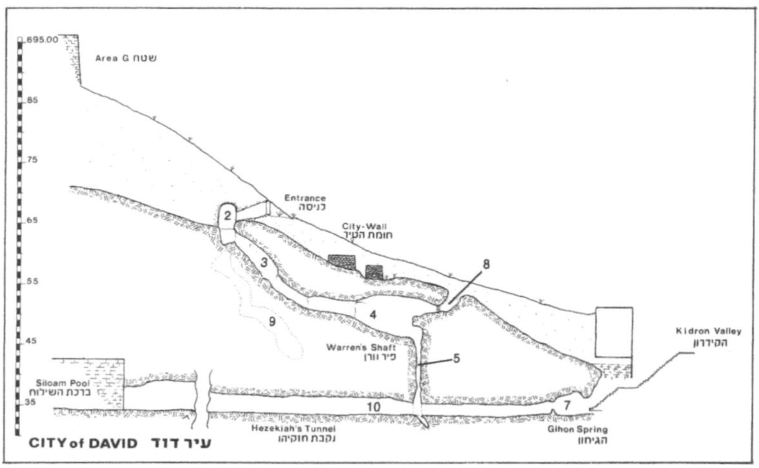

(b)

Fig. 11.1 (**a**) Hezekiah's tunnel and other water systems in Jerusalem in the City of David (Cahill and Tarler, 1994). (**b**) City of David: Plan and section of tunnels in Warren's Shaft: 1. Access tunnel; 2. Vaulted chamber; 3. Stepped tunnel; 4. Horizontal tunnel; 5. Vertical shaft; 6.tunnel linking up to spring; 7. Gihon Spring; side cave; trial shaft; 10. Hezekiah's tunnel (Shiloh, 1994)

Gihon Spring without having to leave the fortified city. The systems consisted of an entrance area, a tunnel, a vertical shaft, and a feeder tunnel from the Gihon to the bottom of the shaft.

The Siloam channel was either contemporary with or slightly later in date than Warren's shaft (Cahill and Tarler, 1994; Shiloh, 1994). This channel consisted of both a tunnel and a stone-capped channel that conveyed water from the Gihon Spring along the eastern slope of the City of David to a reservoir in the Tyropoeon Valley.

The channel also released water through openings in its eastern wall to agricultural plots and collected runoff from the slope through openings in the caps.

One biblical account (Chronicles 32:3) discusses King Hezekiah's tunnel, the latest of the ancient water supply systems. During the First Temple, Jerusalem was under military threat from Assyria (2 Kings 20:20; Isaiah 22:11; 2 Chronicles 32-2-4, 30) causing an urban water crisis (Bruins, 2002). Hezekiah's tunnel was built toward the end of the 8th century B.C. (Shiloh, 1994). The main water supply of ancient Jerusalem was the Gihon Spring, located inside a cave in the Kidron Valley (see Fig. 11.1). Water flowed from the spring intermittingly for about 40 minutes every six to eight hours (Bruins, 2002). Because a siege (during the late eighth century B.C.) could cut off the main water supply, which was just outside the city walls, King Hezekiah had a 533 m tunnel (Isar, 1990) dug in the limestone to channel the water underground into the city with the outlet being the Pool of Siloam (see Fig. 11.1). The tunnel was dug from both ends. Ancient Hebrew inscription was engraved in the rock wall near the outlet of the aqueduct tunnel (Bruins, 2002) describing how the dug tunnels met.

During the Hellenistic and Roman periods aqueducts were built to replace the Gihon Spring as ancient Jerusalem's primary source of water supply. The aqueducts brought water from springs in the Judean Hills. Hezekiah's tunnel and the Pool of Siloam continued to function during these periods (Cahill and Tarler, 1994).

Sennacherib built a series of aqueducts to provide water from the Gomel River. To quell the Assyrians (695 B.C.) he razed Babylon and diverted one of the irrigation canals to wash away the ruins. Sargon II (Assyrian) destroyed Armenian irrigation system and flooded their land (720–705 B.C.). Gleick (1990) provides other accounts in the Chronology.

11.3.3 Security and Sustainable Design

The ancients considered security as one of the critical aspects of the design and construction of their water systems. There are many examples such as the Peisistratean aqueduct in Athens. This aqueduct was built underground not exposing the aqueduct. The eleven aqueducts of Rome were almost entirely under the surface ground as listed in Table 11.1. A combination and balance of smaller scale (wells and cisterns) and the larger-scale water supply measures were used by the ancient Minoans, Greeks, and Romans.

The security of water supply is also a modern day concern around the world, from the viewpoint of both an adequate water supply and the possibility of terrorist activity on our water supply systems. The three obvious attributes of a water supply system are: (a) there must be adequate quantities of water on demand; (b) it must be delivered at sufficient pressure; and (c) it must be safe to use. The first two are influenced by physical damage and the third attribute (water quality) is susceptible to physical damage as well as the introduction of microorganisms, toxins, chemicals or radioactive materials. Actions (terrorist activities) can be debilitating for the water supply system. Because water systems are spatially diverse and many of the system components such as tanks and pumps are located in isolated locations, they have

Table 11.1 Rome's Aqueducts (after Fahlbusch, 1987)

Name	Date	Total length (km)	Length under-ground(km)	Origin
Appia	312 B.C.	17.6	16.8	Springs in Anio Valley
Anio Vetus	272	64	63.6	Anio River
Marcia	144–140	91.2	80	Springs in Anio Valley
Tepula	126	18.4	8.4	Springs near Alban Mountains
Julia	33	22.8	12.4	Springs near Alban Mountains
Virgo	19–21	20.8	19.2	Springs in Anio Valley
Alsietina	10–2	32.8	32.4	Lake Alsietinus
Claudia	38–52 A.D.	68.8	53.6	Springs in Anio Valley
Anio Novus	38–52	86.4	72.8	Anio River
Traiana	109–117	59.2	59.2	Springs near Lake Sabatina
Alexandrina	226	22.4	12.8	Springs at Sasso Bello

inherent potential to be vulnerable to a variety of threats that would compromise the delivery of safe water. The areas of vulnerability include: the raw water source, raw water channels and pipelines, raw water reservoirs, water treatment facilities, connections to the water distribution system, pump stations and valves, and finished water tanks and reservoirs.

Since the September 11, 2001 events in the U.S.A., the security of water supply (in particular water distribution systems) has become a major issue of concern. Basically our water supply systems have not been designed and constructed using sustainable design procedures that consider security as one of the design principles, even though the Greeks and Romans dealt with this concern in their water systems. A distribution system of pipelines, pipes, storage tanks, and the appurtenances such as various types of valves, meters, etc. offers the greatest opportunity for terrorism because it is extensive, relatively unprotected and accessible, and often isolated. Not only the physical destruction of a water distribution system's assets of the disruption of water supply is possible as was the case for the Greeks and Romans, but during our modern time we must also be concerned with the possible introduction of biological and/or chemical contaminants and the disruption of the operation (cyber threats) of the systems. Mays (2004) presented details on the threats categorized as physical threats, chemical threats, biological threats, and cyber threats.

11.4 Societies Do Fail and Collapse

11.4.1 Some that Failed or Collapsed

The use of "collapse" or "failure" of a civilization is somewhat misleading. Civilizations never "die" as they are "a complex configuration of institutions built upon a foundation of shared religious, political, and economic ideas and concepts"

(Demarest, 2004). Even with catastrophic events and declines, the elements of a civilization can continue and transform into a new configuration. The western part of the Roman Empire is a good example. It declined over several centuries with periods of revitalization, breaking up into Gothic kingdoms or states that maintained many institutions of ancient Rome. The eastern Roman Empire continued under Byzantium and flourished. Even though the somewhat colorful terms, collapse and failure, are used herein, there is no implication that these civilizations "died."

The Indus Valley civilization (3000–1500 B.C.) is thought to have collapsed because the Indus River shifted its course and the continued salinization. Mohenjo Daro in the Indus Valley (Pakistan) declined after 2000 B.C. possibly due to climate change, river shifts, and water resources management problems. Droughts possibly caused or attributed to the collapse of the Akkadian Empire in Mesopotamia around 2170 B.C. The Minoan civilizations declined, possibly due to an earthquake disrupting the water supply. Petra, the Nabataean civilization capital in southwestern Jordan, collapsed as a result of the disruption of the elaborate water supply system caused by an earthquake in May of A.D. 363. The Romans and other ancient civilizations in the Mediterranean region are discussed later in this chapter.

Several ancient civilizations in Mesoamerica and the southwestern United States also failed and collapsed. Teotihuacan (city of the Gods), Mexico was abandoned mysteriously around A.D. 600 to 700. Teotihuacan was a very impressive civilization which evolved about 40 km north of Mexico City about the same time as Rome. During this time the collapse of civilized life occurred in most of central Mexico. One possible cause was the erosion and desiccation of the region resulting from the destruction of the surrounding forests that were used for the burning of the lime that went into the building of Teotihuacan. The increasing aridity of the climate in Mexico may have been a related factor. The entire edifice of the Teotihuacan state may have perished from the loss of agriculture. Even though the city had no outer defensive walls, it was not an open city easy for hostile outsiders to attack. The collapse of Teotihuacan opened civilized Mexico to nomadic tribes from the north. Human malnourishment has been indicated from skeletal remains. The collapse/abandonment of Xochicalco, most likely, resulted from drought, warfare, and internal political struggles.

Crisis overtook all the classic civilizations of Mesoamerica (including the Mayans), forcing the abandonment of most of the cities. Some anthropologists believe the crisis may have been a lessening of the food supply caused by a drying out of the land and a loss of water sources to the area. Speculation is that in some areas this might have been caused by a climatic shift toward aridness, particularly in the Puuc centers of western Yucatan (Hoddell et al., 1995; Robichaux, 2002). These included Uxmal, Sayil, Labna, and Kabah. It may have happened all over Mexico during the classic period when the deforestation of the valley occurred. Originally there were cedar, cypress, pine, and oak forests; today there are cactus, yucca, agave, and California pepper trees. Such a change in vegetation indicates a significant climate shift.

Other Mesoamerica civilizations that failed and collapsed include the Mayans discussed below. Ancient civilizations in the southwestern United States that failed and collapsed include the Hohokam (300 B.C.–A.D. 1450) and the Chaco Anasazi (A.D. 600–1200) discussed below. The collapse of the Moche IV civilization on the Peruvian coast around A.D. 600; and the collapse of the Tiwanaka civilization in the Andes around A.D. 1100 are two other civilizations in South America.

11.4.2 Diamond's Framework, Mesoamerica and the Southwestern U.S.

A five-point framework for the collapse of societies was proposed by Diamond (2005):

- Damage that people inadvertently inflict on their environment
- Climate change
- Society's responses to its problems
- Hostile neighbors
- Decreased support by friendly neighbors

Essentially the first three relate, directly or indirectly, to water resources.

Other civilizations that collapsed in ancient Mesoamerica, the Mayans, and in the southwestern United States, the Hohokams and the Chaco Anasazis, have interesting histories. The ancient Maya lived in a vast area covering parts of present-day Guatemala, Mexico, Belize, and the western areas of Honduras and El Salvador. Mayans settled in the last millennium B.C. and their civilization flourished until around A.D. 870. The environment that the Mayans lived in was less fragile than that of the semiarid lands where the Anasazi and Hohokam lived.

Centuries before the Spanish arrived, the collapse of many other great Mayan cities occurred within a fairly short time period. Several reasons have emerged as to why these cities collapsed, including overpopulation and the consequential exhaustion of land resources possibly coupled with a prolonged drought. A drought from A.D. 125 until A.D. 250 caused the pre-classic collapse at El Mirador and other locations. A drought around A.D. 600 caused a decline at Tikal and other locations. Around A.D. 760 a drought started that resulted in the Mayan classic collapse in different locations from A.D. 760 to 910.

The soil of the rain forest is actually poor in nutrients so that crops could be grown for only two or three years, then to go fallow for up to 18 years. This required ever-increasing destruction of the rain forest (and animal habitat) to feed a growing population. Other secondary reasons for the collapse include increased warfare, a bloated ruling class requiring more and more support from the working classes, increased sacrifices extending to the lower classes, and possible epidemics. The Maya collapsed as a result of four of the five factors in Diamond's (2005) framework. Trade or cessation of trade with friendly societies was not a factor for the Maya. Water resources sustainability was likely a factor in the collapse of the Maya.

In A.D. 899 a flood caused decentralization and widespread population movement of the Hohokams from the Salt-Gila River Basin to areas where they had to rely upon dry farming. The dry farming provided a more secure subsistence base. Eventual collapse of the Hohokam regional system resulted from a combination of several factors, which included: flooding in the 1080s, hydrologic degradation in the early 1100s, and larger communities forcibly recruiting labor or levying tribute from surrounding populations (Crown and Judge, 1991). In 1358, a major flood ultimately destroyed the canal networks, resulting in the depopulation of the Hohokam area. Culturally drained, the Hohokam faced obliteration in about 1450. Parts of the irrigation system had been in service for almost 1500 years, which may have fallen into disrepair, canals silted in need of extensive maintenance, and problems with salt.

Chaco Canyon, situated in the San Juan Basin in northwestern New Mexico, had limited surface water, most of which was discharged from ephemeral washes and arroyos. The Chacoans developed a method of collecting and diverting runoff as discussed in Chapter 9. The diversion of water from the mesas (see Fig. 11.2) into the canals combined with the clearing of vegetation resulted in the eroding (cutting) of deep arroyos to depths below the fields being irrigated. By A.D. 1000 the forests of pinon and juniper trees had been deforested completely to build roofs, and even today the area remains deforested as shown in Fig. 11.2. Between A.D. 1125 and 1180, very little rain fell in the region. After 1180, rainfall briefly returned to normal. Another drought occurred from 1270 to 1274, followed by a period of normal rainfall. In 1275, yet another drought began which lasted 14 years.

Of the five-factor framework for social collapse suggested by Diamond (2005), the only factor that did not play a role in the collapse of the Anasazi was hostile neighbors. Water resources sustainability was affected by the deforestation, the erosion (cutting) of the arroyos from the diversion of water resulting in lowering the groundwater levels and the supply source to the irrigated fields, and finally, the repeated periods of drought caused the final collapse.

What relevance does the collapse or decline of ancient civilizations have upon modern societies? Learning from the past and discovering the reasons for the

Fig. 11.2 Chaco Canyon (copyright with L.W. Mays)

success and failure of other societies seems very logical. We certainly are a much more advanced society than those of the ancient societies, but will we be able to overcome the obstacles to survival before us? The collapse of some civilizations may have been the result of the very processes that had been responsible for their success (e.g. the Mayans and Romans and others).

11.5 The Ancient Egyptian Civilization Never Lost Sight of the Past

The predictability of the Nile River floods and the production of grains suggest order and stability. Throughout history the advancements of irrigation of the Nile, starting from natural irrigation and advancing to artificial irrigation and then the development of lift irrigation with the shaduf and then the Archimedes screw (or tanbur), and the saqiya (or waterwheel). From a water management perspective, all evidence known suggests that flood control and irrigation, at the social and administrative levels, were managed locally by the rural population within a basin. The rise and sustainability of Egypt, with so many great achievements, was based primarily on the cultivating of grain on the Nile River floodplain, without a centralized management of irrigation. What is so unique is that Egypt probably survived for so long because production did not depend on a centralized state. Collapses of the government and changes of dynasties did not undermine irrigation and agricultural production on the local level.

Ancient Egyptians lived in harmony with nature for thousands of years. Harmony with nature continued until the construction of the Aswan High Dam, which significantly changed the hydrologic regime of the Nile River Basin in Egypt (Fahlbusch, 2004). The modern Aswan High Dam is one of the most controversial of the existing big dams in the world. Economic benefits of the dam have never been in doubt and the dam has been important in Egypt's economic survival. However, the construction of the dam was accompanied with many side effects that are still controversial. These side effects include channel degradation, silt deprivation, dune accumulation, coastal erosion, increased use of pesticides and fertilizers, rise of water table, problems of drainage, and changes in water quality. These were problems never faced by the ancient Egyptians. The dam prevents sediment from flowing downstream to the fields and to the Mediterranean Sea as was the natural course. Changes in water quality downstream from the dam include the drop in turbidity, increase in total dissolved solids, higher count of undesirable algae, taste problems, increased density of phytoplankton (Said, 1993). The downstream river is becoming a receptacle of domestic, industrial, and agricultural wastes, with conditions in the delta being even worse because of the reduced velocity of the river, concentration of industrial plants and more intense agriculture. Deterioration of the river has affected the fish population in the downstream river. The real question is, are these mega projects with their environmental consequences sustainable for the future?

11.6 Ancient Lessons for Modern Times: The Ancient Greeks

Koutsoyiannis et al. (2008) explored the legacies and lessons on urban water management learned from the ancient Greeks. They summarized the lessons learned:

- The meaning of sustainability in modern times should be re-evaluated in light of ancient public works and management practices.
- Technological developments based on sound engineering principles can have extended useful lives.
- Security, with respect to water, is of critical importance in the sustainability of a population.
- In water-short areas, development of an effective water resources management program is essential.

Democracy was established in Athens around (510–508 B.C.) and lasted up to 322 B.C. with some intervals of oligarchic governance. During that period the city grew and the need for water increased. Because no major water supply projects were constructed, the Athenians built smaller-scale projects, such as cisterns to complement the wells or replaced them as groundwater levels were decreasing. They also developed an institutional framework for sustainable management of water. Public officials were appointed to operate and maintain the city water system, monitor enforcement of regulations, and to ensure fair distribution of water (Koutsoyiannis et al., 2008).

11.7 Ancient Water Technology for Sustainability

The Romans used water as a matter of luxury and prestige building mega water projects using aqueducts to transfer water to their public fountains and baths. During this time many of the smaller water projects such as wells and cisterns were abandoned. This signified a different water management attitude than the classical Greeks. The ancients showed us that it was feasible to construct technologically advanced water transportation projects on a large scale. To name just a few: ancient Greeks built the Peisistratean aqueduct for Athens and the aqueduct of Samos (tunnel of Eupalinos), the Roman aqueduct systems to supply water to Rome, Lugdunum (Lyon, France), Emerita Augusta (Merida, Spain), Serino aqueduct, and many others. I think of these as the ancient mega water supply projects. Were these mega water supply projects the answer and are the present day mega water projects the answer? How de we balance the mega water projects with the methods of traditional knowledge?

11.7.1 Traditional Knowledge

The United Nations organized the World Conference on the Environment and Development held in Rio de Janeiro in 1992. This conference, typically referred to as "The Earth Summit" was aimed at reconciling the dramatic world environmental conditions affecting the development and welfare of people. Three conventions for climate, biodiversity, and desertification were considered from development and technology perspectives with the consideration of traditional knowledge and practices. The United Nations Convention to Combat Desertification (UNCCD) selected a Science and Technology Committee to look at the inventory and classification of traditional knowledge. This effort researched approximately 200 member countries. The following definition of traditional knowledge was developed:

> Traditional knowledge consists of practical (instrumental) and normative knowledge concerning the ecological, socio-economic and cultural environment. Traditional knowledge originates from people and is transmitted to people by recognizable and experienced actors. It is systematic (inter-sector and holistic), experimental (empirical and practical), handed down from generation to generation and culturally enhanced. Such a kind of knowledge supports diversity and enhances and reproduces local resources.

A list of 78 techniques and practices were developed by the committee and were classified into the seven topics: water management for conservation, improvement of soil fertility, protection of vegetation, fight against wind or water erosion, silviculture, social organization, and architecture and energy. Desertification was defined as "deterioration of the land in the arid, semiarid and semi humid dry areas due to factors including climate changes and human activity."

"Modern technology aims at an immediate efficiency through a high specialization of knowledge supported by dominant structures able to mobilize resources external to the environment" (Laureano, 2001). An example of modern technology would be to dig deep wells and pump to an extent that would harm water supplies for the future, which has been done in so many places in both the developing and developed parts of the world. Traditional knowledge would have relied on a system for harvesting meteoric water or exploiting run-off areas using the force of gravity or water catchment methods that would allow the replenishment and increasing the durability of the resource (Laureano, 2001). "Modern technological methods operate by separating and specializing, whereas traditional knowledge operates by connecting and integrating."

The use of traditional knowledge does not directly apply techniques of the past but instead, "to understand the logic of this model of knowledge" (Laureano, 2006). Traditional knowledge allowed ancient societies to keep ecosystems in balance, carry out outstanding technical, artistic, and architectural work that has been universally admired. The use of traditional knowledge has been able to renew and adapt itself. Traditional knowledge incorporates innovation in a dynamic fashion, subject to the test of a long term, achieving local and environmental sustainability.

Most of the drylands are located in developing countries, and therefore the application and development of traditional knowledge and technologies should be encouraged. The present trends are such that the use of traditional water management technologies is falling by the wayside. Adeel (2009) presents four major reasons for this.

- Changes in socio-economics of developing countries have meant that there are fewer skilled experts to develop and manage the traditional systems. These systems are primarily rural, therefore with the rural-to-urban-exodus of people the manpower to maintain them becomes limited. In addition there are few economic incentives for the young to use these systems.
- Newer methodologies such as electrical powered tube wells and irrigation systems that deliver water through canals and channels have greater appeal. Unfortunately these newer approaches are not economically favorable to the traditional approaches when all of the costs are included.
- Very limited research has been performed to improve traditional technologies for coping with population growth, desertification, and other external social and economic driving forces.
- At the international level investment into research, evaluation, maintenance and deployment of traditional technologies has been basically non-existent.

Because of the potential for sustainable management of water resources we could argue that there should be a strategic and pro-active investment in these technologies.

11.7.2 Sassi of Matera: Example of Traditional Use of Water Resources

Water saving techniques such as condensation caves and pits, stone arrangements for rainfall harvesting, and underground dams are fairly well widespread in the Mediterranean region arid and semi-arid areas. Thanks to the ancients these technologies became widespread. In southern Italy, the Sassi of Matera represents a typical example of the traditional use of water resources in the Mediterranean region. The local knowledge system has towns built along the borders of deep valleys (gravines) that have little or no water carrying capacity. Settlements are not placed in the canyon bottoms, but instead were placed on the upper part along the plateau and the steep slopes. The water resources are rainfall and dew that are harvested in drains and cave dwellings. Original prehistoric methods used water production techniques such as catchment, distillation, and condensation, all of which, are used in the Sassi of Matera. During rainfalls terracing and the water collection system protect the slopes from erosion, and water flows by gravity to cisterns in the caves. During dry periods, moisture condenses from the air at night in a final underground cistern.

The town is built as a vertical structure with up to ten levels of caves, one on top of the others, with dozens of bell-shaped cisterns connected to each other by means of canals and water filter systems as shown in Fig. 11.3. Vertical structure of the town allows the use of gravity for water distribution. The town has a maze of small streets, stairs, and underground passage ways that follow the ancient hydraulic structures.

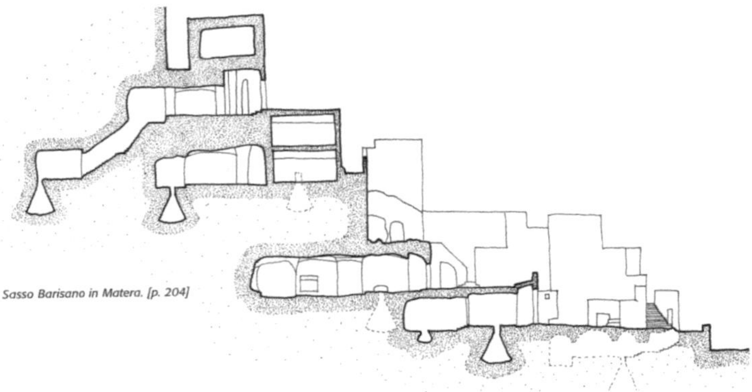

Fig. 11.3 The Sasso of Barisano in Matera (Laureano, 2001)

During the 1950s, the Sassi of Matera was closed as a result of the neglected condition, requiring 20,000 inhabitants to move out. Abandoned houses became the property of the state, who built a wall to prevent reoccupation of the dwellings. In 1986, the Italian government, motivated by individuals involved in cultural activities, allocated money (100 billion liras) to restore the sassi. In 1993 Sassi of Matera became a UNESCO World Heritage Site. By year 2006 around 3000 inhabitants returned to live in the cave homes (Laureano, 2006).

Sassi of Matera is an excellent example of the restoration of traditional water collection systems, illustrating natural resources management capabilities (water, sun, and energy) that were once utilized, but today are neglected. Laureano (2006) is right on target to say that, "It is necessary to maximize the potential of a town at a local level to assure its harmonious and sustainable development." Unfortunately "modernization" often destroys traditional methods, threatening the ecological equilibrium.

11.8 Rome's Water System: Contribution to Success and Failure

11.8.1 The Aqueduct System

Romes's aqueducts were built using war booty and taxing those conquered. The main function of the aqueduct was to supply water for the extravagant lifestyles of the rich, mainly supplying water for the bathes, the fountains and to provide water for private houses of the rich. Certainly there were some benefits for the not so rich, but were probably minor compared to those for the rich.

Prior to the construction of the first aqueduct (Aqua Appia, named after the censor Appius Claudis) to Rome, completed in 312 B.C., the water supply consisted mainly of rainwater, spring water, well water, and river water. Possibly because the Tiber River carried a large amount of sediment throughout the year, it was not widely used for water supply. At this time Rome was a small city located 24 km from the sea, allowing access for trade, without vulnerability from attack by invaders. The amazing thing is that Rome was barely standing on its own at that time, but yet the Appia was constructed to bring additional water to the city. The aqueduct was 17.6 km long, with 16.8 km below the surface of the ground. This was a very large and demanding undertaking that must have been accomplished by people with a great deal of initiative. Obviously the people must have had a firm belief in the future.

Rome's aqueduct system evolved in a piecemeal fashion over a 500-year time period. Table 11.1 lists the eleven aqueducts in Rome along with other information such as the date the aqueduct was built, its length, and it origin. As pointed out by Evans (1994) throughout the history of Rome, aqueduct construction generally was not planned in an orderly manner. During the Republican Rome the government tended to allow needs to become critical before aqueducts were built, a similarity to modern times. Another factor is that the technology of the aqueduct did not seem to change over the time period of over 500 years. One might question, why did the Romans stop building new and better aqueducts and why did they not perfect their water systems? Figure 11.4 shows the views of the some of the aqueducts of Rome at the Porta Maggiore (double walled gate) on the Aurelian wall where all

(a) (b)

Fig. 11.4 Aqueducts at Porta Maggiore on the Aurelian Wall where the aqueducts entered ancient Rome (**a**) Aqueducts Claudia (*top*) and Anio Novus (*bottom*) (**b**) Three aqueducts Julia (*top*), Tepula (*center*), and Marcia (*bottom*) located on the Aurelia Wall (photos copyright with L.W. Mays). Color version available in Appendix

Fig. 11.5 Branch Aqueduct (Aqua Claudia) that supplies the Trophies of Marius nymphaeum (copyright with L.W. Mays)

the eastern aqueducts entered Rome. Figure 11.5 shows the branch aqueduct (Aqua Claudia) that supplies the Trophies of Marius nymphaeum near Porta Maggiore.

11.8.2 Observations of Frontinus

We can learn a lot from reading the works by Pollio Vitruvius (*De Architectura*) and Sextus Julius Frontinus (*De aquaeductu urbis Rome*) who lived 40–103 A.D. So during his writings the first nine aqueducts built in Rome were still being used. Frontinus was a retired military officer who in A.D. 97 took over as director of the Rome Metropolitan Waterworks. Frontinus's work reflects the first nine aqueducts to Rome which existed during his time. So when he became the director the oldest aqueduct (Appia) was 215 years old and the newest was 45 years old. By any modern standards most of these aqueducts would be very old structures. Water loss due to theft and official negligence was a major concern. Water was stolen through illegal taps using branching pipes for illegal diversions. Large amounts of water were also lost from leaks in the aqueducts.

Frontinus was responsible for locating the illegal taps and repairing the leaks. He wrote that the recovered amount of water was virtually the same as finding a new source of water (Frontinus, translation by Charles Bennett, 1925). He also wrote (Book I.7):

> One hundred and twenty-seven years later, that is in the six hundred and eighth year from the founding of the City, in the consulship of Servius Sulpicius Galba and Lucius Aurelius Cotta, when the conduits of Appia and Old Anio had become leaky by reason of age, and water was also being diverted from them unlawfully by individuals, the Senate commissioned Marcius, who at that time administered the law as a praetor between citizens, to reclaim and repair these conduits; and since the growth of the city was seen to demand a more bountiful supply of water, the same man was charged by the Senate to bring into the

> City other waters so far as he could. He restored the old channels and brought in a third supply, more wholesome than these which is called Marcia after the man who introduced it. We read in Fenestella, that 180,000,000 sesterces were granted to Marcius for these works, and since the term of his praetorship was not sufficient for the completion of the enterprise, it was extended for a second year. . . .

Several thoughts come to mind in reading this. Because of the age of the aqueducts Appia and Anio Vetus and the lack of maintenance they leaked extensively and needed repair. One might ask how many of our present day water supply structures will be in use in the next 150 plus years?

According to Frontinus (Book I.9) by the time Julia was constructed the aqueducts of Appia, Anio Vetus (Old Anio), and Marcia had almost worn out and Tepula so Agrippa restored these. Julia had branch pipes by which the water was "secretly plundered." By destruction of the branch pipes the aqueduct maintained its "normal quantity even in times of most extraordinary drought."

Frontinus (Book I.120) described that the necessity of repairs to the aqueducts was the result of damage "by the lawlessness of abutting proprietors, by age, violent storms, or by defects in the original construction, which has happened quite frequently in the case of recent works." Frontinus (Book I.121) recognized the troublesome aspects of the aqueduct:

> As a rule, those parts of the aqueducts which are carried on arches are placed on sidehills and, of those on arches, the parts that cross rivers suffer most from the effects of age or of violent storms. These, therefore, must be put in order with care and dispatch. The underground portions, not being subjected to either heat or frost, are less liable to injury. Defects are either of the sort that can be remedied without stopping the flow of the water, or such as cannot be made without diverting the flow, as, for example, those which have to be made in the channel itself.

Frontinus (Book I.122) discussed the accumulation of deposits and the damage to the concrete linings of the aqueducts. He discussed the fact that repairs to the channels should not be made during the summers, but should be during the spring and autumn and only a single aqueduct should be considered at a time because of the demands. Frontinus (Book II.120) wrote that the Aqua Claudia was badly built. This aqueduct had to be rebuilt and restored many times. In the end it had become practically a different aqueduct.

The strength and efficiency of the Roman aqueduct system directly affects the strength and efficiency of the Roman people and their government (Tradieu, 1986). But then as the Roman government declined so did the water supply system. One might conclude that one of the major accomplishments that helped to make them great also was a factor in their decline. Rome's aqueduct system, by modern day standards, was defective in that it leaked significantly as evidenced by the large deposits of lime and most likely during many times the aqueducts did not function properly. Unfortunately Rome's technology of building aqueducts did not change over time.

11.8.3 Unsustainability of Rome's Water Supply System

Aqueducts were financed from the spoils of various wars and the heavy tribute charged to conquered countries. Rome used the valuable possessions and resources with monetary value from the conquered territories. The economy was based on the spoils of war so that when the chronic state of wars ended and/or the conquered had no resources, the emperor took over the responsibility of public works, obscuring the problem of paying for the construction of the aqueducts. The emperors paid for the large projects by taxing the citizens rather than making the water supply systems profitable. Augustus started the practice of charitable gifts from the private citizens which led to absorption of private capital into public projects without hope of financial returns. None of the investments in any of the aqueducts were ever paid back, only consuming capital with no returns (Tradieu, 1986). The aqueducts were very expensive to build and maintain. In the beginning the government had the money, but as time went on the aqueducts required more money, more maintenance, and more bureaucracy. Eventually the government could no longer afford them and had to rely on taxes. The government's early approach did not consider what the later sacrifices would be for the future generations, no longer having the monies from conquered states to pay for the water systems. The Roman government paid extensively for its good intentions, which turned out to be very unsustainable. Sounds somewhat like our modern day adventures such as senseless wars and extravagant lifestyles, are having on our modern day events around the World.

11.9 The Failure of Angkor: An Ancient Megacity

The ancient city of Angkor, Cambodia (of the Khmer culture) covered more than 160 km^2 in northern Cambodia, situated on the edge of the Great Lake (Tonle Sap). The Classic Angkor civilization was part of the Khmer culture (between A.D. 802 and 1327). Prior to A.D. 802 the Khmer political landscape consisted of a number of independent kingdoms (Coe, 2003). Angkor became the imperial capital of the Khmer Empire. Ancient Angkor was a vast complex of temples built from the 8th to the 13th centuries A.D. Figure 11.6 is map of Angkor showing surface features such as topography and waterways. A study by Evans et al. (2007) concluded that the area of Angkor's urban complex was roughly 900–1,100 km^2 which is almost four times the size of present day New York City. Angkor was a low density city with dwellings and water tanks spread over the area and connected by roads.

Angkor is located in the Lower Mekong Basin which is subject to an annual cycle of monsoons causing alternation between a wet rainy season (summer monsoon) and a strongly marked dry season. The heavy rainfall during the summer monsoon causes the Mekong River and its tributaries to rise and flood low-lying areas. Snow melt in Southwestern China and Tibet flowing down the Mekong contribute to the flood volume. The Tonle Sap River, a tributary of the Mekong, reverses flow because of the back water effects from the large flows in the Delta of the Mekong and causes the water levels in the Great Lake (Tonle Sap) to rise. Floods subside during the

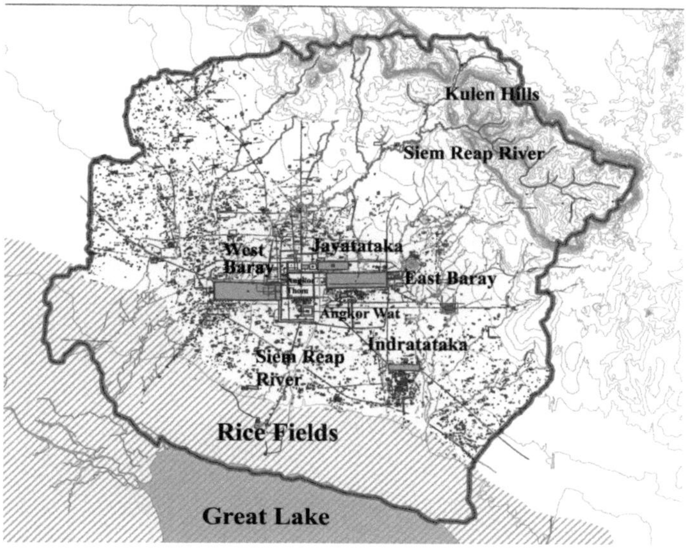

Fig. 11.6 This is a map derived from airborne data sources and earlier archaeological maps. Source: http://earthobservatory.nasa.gov (courtesy of NASA)

winter monsoon and again river flow is toward the Delta causing the water levels in the Great Lake (Tonle Sap) to lower. Total rainfall in the Lower Mekong Basin fluctuates from year to year and is never very high, with an average of 150 cm per year in area of Angkor. In Phnom Penh the mean rainfall is 143 cm and can be as high as 231 cm and as low as 97 cm.

Angkor Wat, a temple complex, is surrounded by a 200 m wide moat (see Fig. 11.7) that measures approximately 1,500 m on each side (a square of approximately 21 ha) and a laterite wall (Coe, 2003). To the north of Angkor Wat is Angkor Thom (Great Angkor), another larger temple complex and location of the Royal Palace, which was a large square 3 km on each side with an area. Angkor is also surrounded with a moat (see Fig. 11.7) and a laterite wall.

The ability to store water was accomplished by constructing large reservoirs called barays. These reservoirs had inlet and outlet control structures so that they were used both in the time of drought and flooding. There were four large barays which had the respective approximate storage volumes (Coe, 2003): West Baray (48 million m^3), East Baray (37.2 million m^3), Jayatataka Baray (8.7 million m^3), and Indratataka Baray (7.5 million m^3). The approximate surface areas of these barays are West Baray (16 million m^2), East Baray (12.4 million m^2), Jayatataka Baray (2.9 million m^2), and Indratataka bay (2.5 million m^2). The West Baray even holds water today. All of these barays may not have been functional at the same time, but one thing is for certain the water management system including the borays and other water infrastructure such as moats, canals, etc. required constant maintenance. A vast canal system was built that was used for both irrigation and transportation.

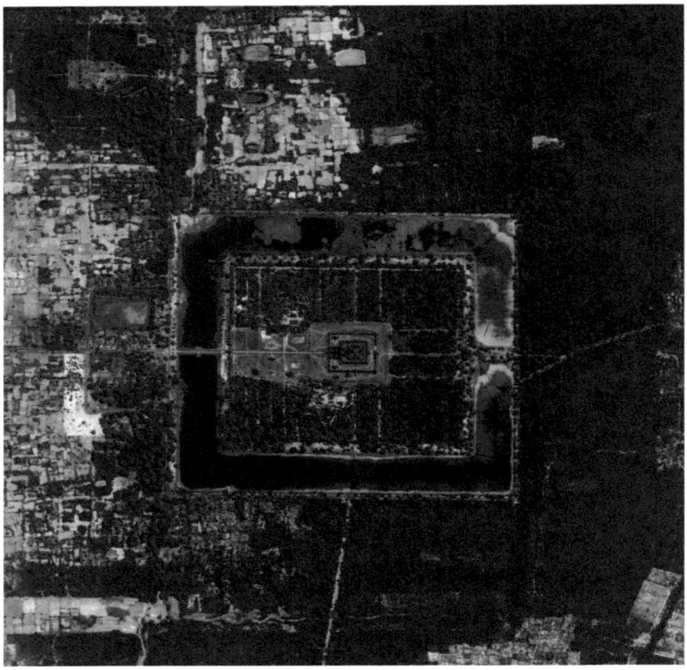

Fig. 11.7 Angkor Wat illustrating the moat surrounding the temple complex (courtesy of NASA)

Angkor was an agrarian state which had rice production as its basic foundation for the classic Khmer civilization; therefore, it would make sense that the most likely purpose of the barays was for providing irrigation water. Borays were built to store water above ground, and were constructed using elevated, earthen dikes. They most likely had other uses such as flood control, recreation, worshiping, and others. In fact Bernard-Philippe Groslier (1979) viewed Angkor as a "hydraulic city". Based upon his calculations the hydraulic system might have supported 600,000 inhabitants. In addition he considered an estimated 429,000 persons that would have been fed by flood-retreat agriculture and another 872,000 people supported by dry rice farming. This would approximate Angkor as having around 1.9 million inhabitants, certainly a mega city of the time. Others such as Acker (1998) concluded that only a small percentage of Grosslier's estimate could have been supported. However, a more recent study by Evans et al. (2007) added to the Groslier's conclusion of Angkor being a hydraulic city built for irrigation to counter the unpredictable monsoons. This study also has confirmed that the Angkor site was the largest pre-industrial city known so far. It was several times the size of Tikal, the Mayan city in Belize.

Early in the 15th century A.D., the Khmer empire basically had ceased to exit. Several theories have evolved as to the downfall of the classic Khmer civilization of Angkor. Grosslier advocated that Angkor fell mainly because of the decline and eventual abandonment of the massive water-management system. He argued that as the system declined it was neglected and filled with silt. Through the use of radar images it has become apparent that rice fields extended from the relatively flat lands of Angkor north to the foothills as a result of increasing population pressure. The lower slopes (Kulen Hills) of the forest cover were logged to clear for cultivation, and also for timber and firewood. This deforestation exposed the complex system of barays and canals to both extensive siltation and flooding and erosion. Probably the barays and canals allowed two annual rice harvests by irrigation and one crop-a-year by flood-retreat farming along the Great Lake. Evans et al. (2007) suggested causes for the failure of Angkor because of the attempt to overcome nature through engineered changes. Deforestation, degradation of top-soil, and erosion, combined with water management operation problems, breaches and failures within the management system contributed significantly to Angkor's failure.

Here is an example of an ancient mega city failing as a result of poor water management practices. The classic Khmer civilization must not have lived in harmony with nature.

11.10 Conclusions

Fahlbusch (2000, 2004) reminds us of the Subak irrigation society in Bali. This irrigation society has over a 1,000 year tradition, with rules based on the philosophy of the "Tri Hita Karana" characterized by three harmonious relationships between: the human being and God, the human beings and the environment, and the human beings themselves. The ancients for the most part lived in harmony with nature and their environment. Those that did not failed. Their actions should be warnings to us, in other words the ancients have warned us. Today we do not live in harmony with nature and the environment. We have created non harmonious relationships among human beings. Instead we continually try to defy nature losing our ability to recognize the warnings of our environment as well as society. We build the modern day mega projects, we pollute, we create problems for our future generations such as our contributions to climate change and our greed for more. Our generation is not sustainable in that it is ignoring the future generations.

References

Acker R (1998) New geographical tests of the hydraulic thesis at Angkor. South East Asia Res 6(1):5–47

Adeel Z (2009) Promoting traditional water management in drylands: adapting traditional knowledge to meet today's challenges. In Adeel Z, Schuster B, Bigas H (eds) What makes traditional technologies tick? A review of traditional approaches for water management in drylands, UNU Desertification Series No. 8, United Nations University, International Network on Water, Environment and health

Bruins HJ (2002) Israel: Urban water infrastructure in the desert. In: Mays LW (ed) Urban water supply handbook. McGraw-Hill, New York

Cahill JM, Tarler D (1994) Excavations directed by Yigal Shiloh at the City of David, 1978–1985. In: Geva H (ed) Ancient Jerusalem revealed. Israel Exploration Society, Jerusalem, pp. 31–45
Coe MD (2003) Angkor and the Khmer civilization. Thames and Hudson, New York
Crown PL, Judge WJ (eds) (1991) Chaco & Hohokam: Prehistoric regional systems in the American southwest. School of American Research Press, Santa Fe, New Mexico
Demarest A (2004) Ancient Maya: The rise and fall of a rainforest civilization. Cambridge University Press
Diamond J (2005) Collapse: How societies choose to fall or succeed. Viking, New York
Evans D, Pottier C, Fletcher R, Hensley S, Tapley I, Milne A, Barbetti M (2007) A comprehensive archaeological map of the world's largest preindustrial complex at Angkor, Cambodia, Proc Natl Acad Sci 104(36):14277–14282
Evans HB (1994) Water distribution in ancient Rome: The evidence of Frontinus. University of Michigan Press, Ann Arbor, MI
Fahlbusch H (1987) Vitruvius and Frontinus – Hydraulics in the Roman period. In: Garbrecht G, Balkema A.A (eds) Hydraulics and hydraulic research: A historical review. Boston, MA
Fahlbusch H (2000) Water in human life: Technical innovations in hydraulic engineering in the last 5000 years, ICID J 49(4)
Fahlbusch H (2004) Men of dikes and canals, in Men of dikes and canals: The archaeology of water in the Middle East, Eds. Hans-Dieter Bienert and Jutta Haser, (International Symposium held at Petra, Wadi Musa (H.K. of Jordan) 15–20 June 1999), VerlagMarie Marie Leidorf GmbH and Rahden/Westf
Frontinus, Sextus Julius (1925) The aqueducts of Rome, translated by Charles Bennett
Gleick PH (2005) The world's water, The biennial report on freshwater resources. Island Press, Washington, DC
Gleick, PH, Loh P, Gomez S, Morrison J (1995) California water 2020: A sustainable vision. Pacific Institute for Studies in Development, Environment and Security, Oakland, CA
Hoddell DA, Curtis JH, Brenner M (1995) Possible role of climate in the collapse of classic Maya civilization. Nature 75:391–394
Isar AS (1990) Water shall flow from the rock: hydrogeology and climate in the land of the bible. Springer-Verlag, Berlin
Koutsoyiannis D, Zarkakoulas N, Angelakis AN, Tchobanoglous G (2008) Urban water management in ancient Greece: Legacies and lessons. J Water Res Planning Manage, ASCE, 134(1)
Laureano P (2001) The water atlas: traditional knowledge to combat desertification. Copyight by Peatro Laureano, Translated by Anna Cirella and Angela Whitehouse
Laureano P (2006) Ancient water techniques for proper management of Mediterranean ecosystems, Proceedings, 1st IWA International Symposium on Water and Wastewater Technologies in Ancient Civilizations, ed. by A.N. Angelakis and D. Koutsoyriannis, National Agricultural Research Foundation, Iraklio, Greece, 28–30 Oct 2006
Mays LW (ed) (2004) Water supply security systems. McGraw-Hill, New York
Mays LW (ed) (2007) Water resources sustainability. McGraw-Hill, New York
Robichaux, MH (2002) On the compatibility of epigraphic and archaeological data, with a drought-based explanation for the classic Maya collapse. Ancient Mesoamerica 13: 341–345
Said R (1993) The river Nile: Geology, hydrology and utilization. Pergamon Press, New York
Shiloh Y (1994) The rediscovery of the ancient water system known as "Warren's Shaft." In: Geva H (ed) Ancient Jerusalem revealed, Israel Exploration Society, Jerusalem, pp. 46–54
Tradieu G (1986) The Roman water system: its contribution to the success and failure of ancient Rome, M.A. Thesis, Department of History, San Jose State University, San Jose, CA

Bibliography

American Society of Civil Engineers (ASCE), Task Committee on Sustainability Criteria (1998) Sustainability criteria for water resources systems, American Society of Civil Engineers, Reston, VA

Brundtland Commission (1987) Our common future, WCED

Coopey R, Fahlbusch H, Hatcho N, Jansky L (2005) A history of water issues: lessons to learn. United Nations University, Toyko, Japan

Groslier B-P, Arthaud J (1957) The arts and civilization of Angkor. Praeger, New York

Laureano P (1998) Proper uses of natural resources, Environmental architecture and hydraulic technologies for self-sustainable and resources-sparing projects. Human Evolution 13(1)

Laureano P (2009) Ancient water management techniques to contrast drought and desertification in the Mediterranean, 2nd IWA Specialty Conference, Bari, Italy, May 2009

Marshall J (2005) Megacity, mega mess .... Nature 437:312–314

Mays LW (ed) (2008) Integrated urban water management: Arid and semi-arid (ASA) regions, UNESCO-IHP, Paris, France, Taylor and Francis, London

Morgan MH (translator) (1914) Vitrivius, The ten books on architecture. Cambridge, MA

National Geographic (2009) Angkor: Why an Ancient Civilization Collapsed, July, pp. 36–55

Stone R (2006) The end of Angkor. Science 311(5766):1364–1368

Unver O (2007) Water-based sustainable integrated regional development. In: Mays LW (ed) Water resources sustainability. McGraw-Hill, New York, pp. 235–265

Vivian RG (1990) The Chacoan prehistory of the San Juan Basin. Academic Press, San Diego, CA

World Water Assessment Programme (2003) Water for people, water for life, The United Nations World Water Development Report 1, UNESCO Publishing, Paris

World Water Assessment Programme (2006) Water, a shared responsibility, The United Nations World Water Development Report 2, UNESCO Publishing, Paris

World Water Assessment Programme (2009) Water in a changing world, The United Nations World Water Development Report 3, UNESCO Publishing: Paris and Earthscan, London

Appendix

(a)

(b)

Fig. 1.5 Water components in Knossos. (**a**) Pipes made of terracotta. (**b**) Drainage channel. (**c**) Carved stone elements of rainfall harvesting system collecting water falling from roof. (**d**) Stepped water channel and sedimentation (desilting basin). Along the stairway is a small channel (for rainwater collection) consisting of a series parabolic-shaped stepped chutes that convey rainwater down steam to the sedimentation tank or basin (copyright permission with L.W. Mays)

L.W. Mays (ed.), *Ancient Water Technologies*, DOI 10.1007/978-90-481-8632-7,

(c)

(d)

Fig. 1.5 (continued)

(a)

(b)

Fig. 1.6 Water components in Phaistos (**a**) Courtyard used for rainfall harvesting with cisterns (*round structures*) shown in background to the right. (**b**) Exit of main drain at southern end of palace (copyright permission with L.W. Mays)

(a)

(b)

Fig. 1.7 Water system in Tylissos, Crete. (**a**) Aqueduct bringing water from springs. (**b**) Sedimentation tank in foreground with stone channel connecting to cistern. (**c**) Sediment tank (**d**) Channel connecting sediment tank to cistern. (**e**) Steps leading down to cistern (photos copyright by L.W. Mays)

(c)

(d)

Fig. 1.7 (continued)

(e)

Fig. 1.7 (continued)

(b)

Fig. 1.8 Peisistratean aqueduct (**b**) Lead pipe joint and elliptical pipe opening for cleaning joint (photos copyright by L.W. Mays)

Fig. 1.10 Inverted siphon at Patara (copyright permission with L.W. Mays)

Fig. 1.13 Tower of the Winds. Located below the Acropolis in Athens, designed by the famous astronomer Andronikos of Kyrrhos to be an elaborate water clock (on the inside). The name is derived from the personifications of the eight winds carved on the eight sides of the building. Originally thought to have been built in the 1st century B.C. during the early Roman Empire, but now many archaeologists think the construction was in the mid-2nd century B.C. during the Hellenistic period (copyright permission with L.W. Mays)

Fig. 4.9 Latrine at the gymnasium of Philippoi. The entrance to the latrine is between the two vertical columns (copyright permission with G. Antoniou)

(a)

Fig. 5.3 (**a**) The outlet of Hasan-Abad subterranean canal is located in Seidan village (**b**) The earth conveyance canal of Hasan-Abad

(b)

Fig. 5.3 (continued)

(a)

Fig. 6.4 Alysia dam: (**a**) General view (**b**) masonry (**c**) spillway

(b)

(c)

Fig. 6.4 (continued)

Fig. 6.7 Roofed cistern (city of Ancient Thera)

(a)

Fig. 7.1 Roman dams near Merida, Spain. (Copyright permission with L. W. Mays)

(b)

Fig. 7.1 (continued)

(c)

Fig. 7.3 Examples of Roman aqueducts (**c**) Aqueduct of the River Gier near Chaponost, France that supplied the Roman city of Lugdunum (Lyon, France). (**d**) Pena Cortada aqueduct near Chelva, Spain. (Copyright permission with L.W. Mays)

(d)

Fig. 7.3 (continued)

(a)

Fig. 7.4 Example Roman aqueduct bridges. (**a**) Pont du Gard aqueduct bridge near Nimes, France. (**b**) Aqueduct bridge of Ferreras (known as Puento del Diablo (devil's bridge) near Tarragona, Spain built during the 1st century B.C. Aqueduct bridge is 217 m long and 27 m high. Bridge consists of two tiers of arches with diameters of 5.9 m (20 Roman feet) with a 15 cm variation and a distance between centers of piers of 7.95 feet (26 Roman feet). (**c**) The *specus* of the aqueduct bridge of Ferreras has a modern day width of 2.5 Roman feet. (Copyright permission with L.W. Mays)

(b)

(c)

Fig. 7.4 (continued)

(a)

Fig. 7.6 *Castellum divisorium* in Pompei. (**a**) Building housing the *castellum*. (Copyright permission with L.W. Mays)

(a)

Fig. 7.7 Components of water distribution system in Pompei. (**a**) Water tower (secondary *castella*) with a storage tank was mounted on top of the tower. Lead pipes were used for the flow of water to and from the tank and were placed in the vertical recessed portion on the tower. The exiting pipes (*calices*) branched off to supply various customers and public fountains. (Copyright permission with L.W. Mays)

(a)

(b)

Fig. 7.8 Roman Cisterns (**a**) Cistern in Illici, Spain (**b**) Cistern below Acropolis in Athens. (Copyright permission with L.W. Mays)

(a)

Fig. 7.11 Perge, Turkey (**a**) Two-story monumental fountain where water overflowed into a canal that divided the colonnaded street. Covered canal delivered water to a pool behind the façade. Water from the pool flowed through an opening just below the reclining statue of the river god Cestrus. (**a**) Fountain and the canal downstream which divided the colonnaded street. (Copyright permission with L.W. Mays)

Fig. 7.12 Remains of monumental fountain in Miletus (Turkey). To the left in back of the fountain is part of the Roman aqueduct that supplied water to the fountain. (Copyright permission with L.W. Mays)

Fig. 7.13 Fountain in ancient Chersonisos, Crete, Greece. Mosaic shows a fisherman and many fish. (Copyright permission with L.W. Mays)

Fig. 7.14 Canopus at Villa Hadrian's Villa (Villa Adriano) at Tivoli, Italy. (Copyright permission with L.W. Mays)

(a)

(b)

Fig. 7.17 Components of the water supply system in Merida (**a**) Only remains of the Las Thomas aqueduct bridge across the Rio Albarregas located near the hippodrome; (**b**) Las Thomas aqueduct downstream of the aqueduct bridge near the Roman theater. Shown on the right side of the aqueduct is a lion head made of stone which was used as a gutter spout. The remains of a castellum is nearby. (**c**) Los Milagros aqueduct bridge of the Proserpina aqueduct across the Rio Albarregas; (**d**) The remains and the reconstruction of the castellum downstream of the Los Milagros bridge; (**e**) Cornalvo aqueduct supplying water from the Cornalvo dam to Merida. Only a few remants of this aqueduct are still visible near the bullfighting arena. (Copyright permission with L.W. Mays)

(c)

(d)

Fig. 7.17 (continued)

(e)

Fig. 7.17 (continued)

(b)

Fig. 7.19 Roman water mill located in the Greek Agora in Athens. This vertical water mill was an overshot type as evidenced by the way in which the lime deposits formed on the walls. (**b**) Wheel pit (or wheel race) for water mill with a branch of Hadrian"s aqueduct supplying water for the mill shown in the background in top of picture. The mill race is shown in the top of the picture. Note the carved area on the right where the horizontal wheel shaft was mounted and the opening for the wheel shaft on the left. The wheel shaft was connected to a vertical cogwheel located outside the wheel pit. (**c**) Water flowed from the wheel pit (wheel race) through the drain on the north wall in the bottom of the pit. Also shown on the east wall are the scratches made by the wheel. (Copyright permission with L.W. Mays)

(c)

Fig. 7.19 (continued)

Fig. 8.2 The Cardo Maximus

Fig. 8.3 Excavations in the North Eastern area of the city

Fig. 8.4 Excavated terracotta pipes used during phase 1

Fig. 8.6 Outer aqueduct delivering water into a cistern

Fig. 8.7 Aqueduct inside the city limits

Fig. 8.9 Outer aqueduct, cistern inside the guard tower, inner aqueduct, first derivation and cistern at the end of the first derivation

Fig. 8.13 End of the first derivation in a cistern

(c)

(d)

Fig. 9.8 Photos of Xochicalco (**c**) Cistern, (**d**) Water pipes (photos copyright by L.W. Mays)

Fig. 9.17 Main aqueduct for Tipon. Note the stair steps on the steep terrain (photo copyright by Yuri Gorokhovich)

Fig. 9.18 Elevated portion of main aqueduct in Tippon (photo copyright by Yuri Gorokhovich)

Fig. 9.19 The source point of the main aqueduct above Tipon (photo copyright by Yuri Gorokhovich)

Fig. 9.20 Restored Inca fountain in Tipon. This is the principal fountain receiving water from the main spring in the background (photo copyright by Yuri Gorokhovich)

Fig. 9.21 Tipon spring located at the head of the central terrace ravine. Water distributed to ceremonial fountains and to irrigated agriculture through canals (photo copyright by Yuri Gorokhovich)

Fig. 9.22 Hydraulic drop structures near the middle of the central terraces in Tipon (photo copyright by Yuri Gorokhovich)

Fig. 9.23 View from Pisac on Vilcanote River and the Yucay Valley (photo copyright by Yuri Gorokhovich)

Fig. 9.24 Terraces in Pisac. Above terraces are living quarters (photo copyright by Yuri Gorokhovich)

Fig. 9.25 Central Inti-Huatana in Pisac bordered by the water channel connected to ritual baths (photo copyright by Yuri Gorokhovich)

Fig. 9.26 The segment of restored channel bordering Inti-Huatana in Pisac (photo copyright by Yuri Gorokhovich)

Fig. 9.28 Tambo Machay complex consisting of retaining wall and calvert with dual channels (*in the center of the picture*) (photo copyright by Yuri Gorokhovich)

Fig. 9.29 Puca Pucara complex (photo copyright by Yuri Gorokhovich)

Fig. 9.30 Dry fountain with dual channels next to Puca Pucara complex. Tambo Machay is located at the upper left corner of the picture (photo copyright by Yuri Gorokhovich)

(a)

Fig. 11.4 Aqueducts at Porta Maggiore on the Aurelian Wall where the aqueducts entered ancient Rome (**a**) Aqueducts Claudia (*top*) and Anio Novus (*bottom*) (**b**) Three aqueducts Julia (*top*), Tepula (*center*), and Marcia (*bottom*) located on the Aurelia Wall (photos copyright with L.W. Mays)

(b)

Fig. 11.4 (continued)

Index

MIX
Papier aus verantwortungsvollen Quellen
Paper from responsible sources
FSC® C105338

If you have any concerns about our products,
you can contact us on
ProductSafety@springernature.com

In case Publisher is established outside the EU,
the EU authorized representative is:
Springer Nature Customer Service Center GmbH
Europaplatz 3, 69115 Heidelberg, Germany

Printed by Libri Plureos GmbH
in Hamburg, Germany